Lynne Malcolm is a science journalist and broadcaster with a deep passion and fascination for people, their behaviour and the workings of the human mind. For nine years, she produced and presented the hugely popular ABC Radio National program and podcast *All in the Mind*, in which she explored a range of topics including neuroscience, psychiatry, psychology, cognitive science, mental health and human behaviour. *All in the Mind* is one of the most downloaded ABC podcasts in Australia and internationally.

Lynne has received numerous media awards and contributes to media award judging panels, including the Australian Museum's Eureka Awards and the Mental Health Service of Australia and New Zealand media awards. She continues to write, podcast and communicate about the topics she's inspired by. *All in the Mind* is her first book. She lives in Sydney.

ALL IN THE
MIND

LYNNE MALCOLM

ABC
BOOKS

The ABC 'Wave' device is a trademark of the
Australian Broadcasting Corporation and is used
under licence by HarperCollins*Publishers* Australia.

HarperCollins*Publishers*
Australia • Brazil • Canada • France • Germany • Holland • India
Italy • Japan • Mexico • New Zealand • Poland • Spain • Sweden
Switzerland • United Kingdom • United States of America

HarperCollins acknowledges the Traditional Custodians
of the land upon which we live and work, and pays respect
to Elders past and present.

First published in Australia in 2023
by HarperCollins*Publishers* Australia Pty Limited
Gadigal Country
Level 19, 201 Elizabeth Street, Sydney NSW 2000
ABN 36 009 913 517
harpercollins.com.au

A catalogue record for this book is available from the National Library of Australia

ISBN 978 0 7333 4242 4 (paperback)
ISBN 978 1 4607 1474 4 (ebook)

Cover design by Louisa Maggio, HarperCollins Design Studio
Author photograph by Jom Photo, jom.com.au
Typeset in Bembo Std by Kelli Lonergan
Printed and bound by CPI Group (UK) Ltd, Croydon, CR0 4YY

This book is dedicated to my dear mum and dad, whom I miss greatly; my son, Sam; my daughter, Eleni; and my partner, Greg — for their loving support.

My deep gratitude goes to the many people who've generously shared their wisdom and personal insights into what makes us exquisitely human.

CONTENTS

INTRODUCTION

As a young child, I worried about walking to school on my own. I worried about my school performance and interactions with my teachers and peers. I worried about people who appeared to be less fortunate than I was. And I had other, free-floating fears about the future. At times this anxiety became overwhelming, so that I couldn't face going to school and putting myself out into the world.

My mother also found it upsetting, but was unsure of how to help. I suspect it reminded her of the anxieties she'd experienced herself at different times of her life. Still, she arranged for me to have some counselling, and it helped me through that difficult period. Of course, since that time I've moved in and out of mental health challenges – seeking help when I felt I needed it, as well as providing understanding and support to my loved ones as they've experienced similar challenges themselves.

On reflection, I think that early experience was the thing that first aroused my fascination with the mind and human behaviour, and set me on a path I continue to travel to this day.

But there would be a few twists and turns in the road before *that* happened.

As a child I was also drawn to the natural world. Our family lived close to the ocean, and I would spend hours exploring the rock platforms and pools, hoping to catch glimpses of fish, octopuses, crabs, starfish and snails. My grandmother used to take my brother and me on walks in nearby bushland, pointing out the native plants and trees that were her passion and are now mine.

Dad was an engineer, and he instilled in me a curiosity about how things worked, an analytical style of thinking and an appreciation for logic. At the same time, my mother was a wonderful storyteller. When she was on a roll, she could really hold the room. She gave *me* a desire to tell stories too.

With parents like mine, you won't be surprised to hear that I did well in English at school, but I was equally strong in science subjects. Fortunately, my selective girls' school actually *encouraged* girls in the direction of a science career if they showed an aptitude.

That childhood fascination with the human mind had never left me, so when it came to thinking about university and a future career, I was intent on becoming a psychologist.

Fast-forward several years, and following uni I opted to travel and see the world. Finally, after some enriching travel and alternative living, it was time to find a career!

I was sitting in the food hall of a shopping centre, browsing the employment section of the newspaper, when an elderly lady sitting next to me peered over my shoulder and said: 'Are you looking for a job? You should try the ABC. I've worked there for years and I love it!'

So I called in to the ABC offices that very day. The following Monday I was offered a temporary job in the Human Resources Department.

It wasn't very stimulating, but I soon discovered that the ABC had a unit set up specifically to train women in broadcasting, called the Australian Women's Broadcasting Co-operative. I began to spend my lunch hours in this women's unit, where I picked up as much as I could from the wonderful, generous women working there. And guess who I ran into? Dodie Coates, the elderly woman from the shopping centre. She was still working enthusiastically in the women's unit as a typist.

After a while, to my delight, I got a job in the unit as Broadcast Production Officer. I was thrown into radio broadcasting at the deep end, and it set me up for an intellectually nurturing lifetime career at ABC Radio National (RN).

A few years later, returning from maternity leave, I joined RN's Science Unit, eventually becoming Executive Producer. Finally I was able to indulge my childhood love of the natural world, and be paid to do it!

In 2012, I was invited to take over *All in the Mind* – a program exploring every aspect of the brain, the mind and human behaviour – which had been hosted for a decade by the amazing Natasha Mitchell. It really was the perfect fit for me, combining my deep curiosity about how the human mind works with my great love of interacting with people.

I took tremendous delight in producing and presenting *All in the Mind* for the following nine years, until I left the ABC at the end of 2020.

Each week, our guests were a mix of scientists, academics, psychologists, psychiatrists, mental health practitioners and those with invaluable first-hand experience. As a producer, my aim was always to begin the program with a personal story if possible, then bring in the perspective of science professionals and academics. I've always believed that this interweaving of personal stories with the science is a powerful way to cast new light on important subjects.

I was always looking for topics I thought would offer the audience new perspectives, and hope within their own lives. So where did I find these topics? I'd read news stories, articles and books. I'd listen to speakers at conferences and public events. Colleagues and friends would make suggestions, and sometimes members of the public would write to me with their ideas.

Until the COVID-19 pandemic, I would interview people in person whenever possible, either in the studio or in their home, place of work or a similar environment. If the guest was interstate or overseas, I'd interview them studio-to-studio on an audio line.

It brought me great joy to converse with scientists, professionals and academics who were dedicated to their important work and keen to share their passion with me. Equally, what a privilege it was to speak to so many brave and inspirational people about their own experiences. I was deeply aware of how difficult it was for them to share their personal stories, but they had decided to take the leap and reveal themselves to me in the hope that others could learn and benefit from their journey. I am deeply grateful to all those who put their trust in me.

It is the incredible conversations I've had with all of these smart, sensitive and brave individuals that form the basis of this book.

* * *

The brain is perhaps the most important organ in our body, and yet in many ways it remains unfathomable to us. It's the source of all our thoughts, feelings, hopes, dreams, memories, values, beliefs, habits and personality traits – everything that makes us who we are – but after millennia of thought and research, we've barely begun to unravel its mysteries.

In the fourth century BC, the Greek philosopher Aristotle thought the brain was simply a secondary organ that kept the heart from overheating, and a place where the spirits circulated freely. Around AD 170, the Roman physician Galen concluded that mental activity occurred not in the heart, as Aristotle had suggested, but in the brain. He posited that the brain's four fluid-filled cavities, or ventricles, were responsible for our personality, memory and other brain functions.

In the 16th century, the Belgian anatomist Andreas Vesalius correctly argued against this idea. We now know that the ventricles are filled with cerebrospinal fluid, which nourishes the brain and protects it against physical impact. In 1791, the Italian scientist Luigi Galvani showed that electricity applied to nerves could make muscles contract – which suggested that electrical impulses were important in the nervous system.

The famous case of 19th-century American railroad worker Phineas Gage contributed to our knowledge about specific parts of the brain. In 1848, Gage had an accident in which an iron spike passed through his left **frontal lobe**. Though he survived, aspects of his personality changed, providing clues to the different functions of individual brain regions. (I'll share a similar but more recent case in Chapter 3.)

By the early 1900s, with the development of microscope technology, anatomists could explore the smallest parts of the brain. The 1906 Nobel Prize in Physiology or Medicine was awarded to neuroanatomists Santiago Ramón y Cajal and Camillo Golgi for identifying nerve cells, or neurons, as the building blocks of the brain. In 1963, Alan Hodgkin, Andrew Huxley and Australian Sir John Eccles won a Nobel Prize in Physiology or Medicine for showing how neurons communicate via electrical and chemical signalling.

Since then, brain research has exploded, with rapid technological advances and scientific collaboration leading to extraordinary leaps in understanding.

According to Professor David Eagleman (whom I'll introduce in Chapter 4): 'We are moving faster and faster for two reasons. The first is that we have this terrific capacity as a human to rest on the shoulders of everything that has come before us ... But the other thing is there are just many more humans now, so we've got almost 8 billion humans running around the planet, all resting on what has come before ... And as soon as someone does something creative it spreads essentially instantly because of the internet and global communication. And so things are just taking off in an enormously rapid way.'

We are learning more and more, not just about how our brains work, but also about all the things that can go wrong. And gradually, the stigmas surrounding mental 'abnormality' are breaking down. We're gaining more understanding of those whose brains are different – whether they want to be accepted for exactly who they are, or are desperate for relief and healing from a debilitating condition or illness.

As findings increase with the assistance of rapidly evolving technology, scientists are developing new and innovative approaches to mental health treatment. Crucially, society is becoming more aware of the need to consider the individual life experiences of people with mental health challenges.

The new global interconnectedness has also encouraged us to look beyond the narrow confines of traditional Western medicine. Many Eastern and indigenous cultures have traditionally viewed the body, mind and spirit as one, and applied this belief to the prevention and healing of disease. Until relatively recently, Western science has seen the body as being entirely separate from emotion and thought, which has reinforced the belief that everything can be cured by drugs and surgery. Things are changing now, with the increase in scientific evidence that the mind can play a role in healing the body. There's a growing realisation that pharmaceutical medications, while life-saving for some people, are inappropriate for others because of intolerable side-effects.

This holistic approach is being applied not just to connect mind and body, but also to connect different fields of learning. We are seeing an emerging appetite for a multidisciplinary approach to mental health issues. Hope for the future rests in the growing knowledge coming from all fields of neuroscience, cognitive science, psychology, philosophy and contemplative thought. One field of study cannot provide all the silver bullets. Human beings work best together, sharing knowledge and resources and collaborating for a better future.

* * *

I believe this is a book that *everyone* needs to read.

Does my brain really have the power to change itself? What exactly is human consciousness? How can I improve my memory, and harness the full power of my creativity? What happens when

I go to sleep, and can my dreams have an impact on how I live my life during the day? More awareness of how our brains and minds work would make all our lives so much richer.

However, many of us are more interested in what happens when they don't work as they should. From 'neurodiverse' conditions like milder autism, to severe mental illness caused by deep trauma, most of us have felt the effects of problems in the brain – whether personally, or through someone we know.

Over the last few years, through the challenges presented by the global COVID-19 pandemic, on top of extreme weather events and increased living costs, we have seen an escalation in the number of people experiencing mental ill-health – some for the first time in their lives. Many others have lived with chronic conditions for years.

Countless families have been affected by the scourge of dementia – including mine, as I'll reveal in Chapter 11. But there are also the conditions that people don't mention. That work colleague who seems to be ignoring you may not be impolite – they may be suffering from an irregularity of the brain. (As I'll explain in Chapter 9, this is exactly what happened when I met radio science guru Dr Karl.)

The first step towards accepting people's differences, or assisting those who suffer, is to understand what they're going through. So come with me now as I share some of the insights I've gained into a range of mental conditions by listening to people who've generously shared their stories – as well as what I've learnt from the scientists and medical practitioners working to make all our lives better by unlocking the seductive mysteries of the brain and mind.

Some points to note:

- Words marked with an asterisk (*) are defined in the glossary at the end of the book.
- References for each chapter can be found at the back of the book.

- A number of the mental health conditions discussed in this book have resulted from experiences of serious trauma. There are sections of the book that include content some readers may find disturbing. This is particularly true of **Part III: New Insights Into Debilitating Mental Conditions.**

Part I

NEW INSIGHTS INTO
HOW THE BRAIN
AND MIND WORK

1

OUR CHANGEABLE BRAIN

Conversations with:

Dr Norman Doidge, psychiatrist and researcher,
University of Toronto

Clark Elliott, brain injury sufferer and Associate Professor
of Artificial Intelligence and Cognitive Science,
DePaul University, Chicago

Deborah Zelinsky, optometrist, the Mind-Eye Institute, Chicago

Abbie Kinniburgh, corpus callosum agenesis sufferer

Maree Maxfield, Abbie's mum

Honorary Professor Linda Richards, corpus callosum researcher,
Chair, Department of Neuroscience and Edison Professor
Neuroscience, Washington University School
of Medicine in St. Louis

The fascinating field of neuroplasticity

John Pepper has Parkinson's disease, a condition caused by damage to the brain cells that help us make automatic movements. This damage leads to difficulties with movement, balance and walking.

John wasn't responding to conventional medication. But one day he decided to join his wife in a get-fit program. It made him pay very close attention to the individual movements involved in walking.

He found he could do these movements with a different level of awareness, using a part of his brain that was not damaged by Parkinson's disease. He was using activity, thought and movement to stimulate dormant circuitry in his brain, which found other ways to overcome the problem. By being conscientious with his walking exercises, he saw remarkable improvement in his symptoms.

* * *

Michael Moskowitz is a physician and psychiatrist who suffered severe chronic pain after a series of accidents.

Michael read the scientific literature and found that there are about a dozen regions in the brain that cause us to feel pain. One of them processes both pain and our ability to visualise.

Michael's solution was to force himself to visualise whenever he was in pain. After several weeks, his pain improved, and after several months, he could go up to 20 minutes without pain. At the end of that year, he felt no pain at all.

* * *

Canadian psychiatrist and researcher **Dr Norman Doidge** is inspired by stories like these. When he released his book *The Brain That Changes Itself* in 2007, it became an international bestseller, and the world learnt a new word: **neuroplasticity.***

Since then, Doidge has continued to explore the powerful potential of this exciting field of research. I spoke to him in 2015, just before he released his second book, *The Brain's Way of Healing*. He told me stories of patients suffering from conditions such as multiple sclerosis, **autism spectrum disorder (ASD)**,* attention deficit hyperactivity disorder (ADHD) and traumatic brain injury. Incredibly, they've all succeeded in healing their brains without either medication or surgery.

Many of the stories of healing Doidge tells may sound almost miraculous to those still wedded, almost dogmatically, to a model that declared the brain can't change under any circumstances. His point is that when we update that old model we see these improvements are not 'miraculous', but, in some cases possible, and they make scientific sense.

How neuroplasticity works

First, a few basics about the brain. It's a bit like a powerful computer that controls our body's functions. Extending from it is the spinal cord, a long bundle of nerves inside the spinal column. Together they make up the **central nervous system**, which allows messages to flow back and forth between the brain and the body.

The central nervous system is made up of millions and millions of microscopic neurons (as well as glial cells, which play a supportive role). Neurons are responsible for receiving sensory input from the world around us, and for sending motor commands to our muscles and the rest of our body. So, for example, if you see a car speeding straight towards you, your nerves and brain communicate so that you can jump out of the way.

When neurons are stimulated, they communicate with other neurons through electrical impulses that cause the release of **neurotransmitters**, chemicals that act as messengers. These neurotransmitters pass through the connection, or synapse, between the sending cell and the receiving cell. This process is repeated from neuron to neuron, making trillions and trillions of connections, and providing a web of fast and efficient communication throughout all parts of the brain and the body that allows us to move, think, feel and communicate.

When you learn new things, for example, the information needs to travel from one neuron to another, over and over. Eventually the brain creates pathways between neurons, so it becomes easier to recall the information and put it to use. If you think about learning to ride a bike, at first it is a challenge to steer, apply the brakes and watch the road all at the same time, but with practice, the neurons create a 'bike riding' pathway.

Early researchers believed that the creation of new neurons stopped shortly after birth. It's now understood that the brain has the incredible ability to establish new connections, reorganise pathways and in some cases even create new neurons.

Canadian neuroscientist Donald Hebb introduced the theory of brain plasticity in his 1949 book *The Organization of Behavior*. It's often summarised in the phrase 'Neurons that fire together wire together.' It means the brain is able to adapt to a changing environment by strengthening new connections and 'pruning away' the weaker ones.

Plasticity continues throughout life, but the young brain changes the most. We can't help but notice how easily young kids seem to pick up a new language, learn to play a musical instrument, or master the ability to recite a comprehensive list of dinosaur species! When a child is born, each of their neurons has around 2500 synapses, but by the time they turn three, this number has increased to 15,000.

This number is halved in the average adult, because as we have new experiences some connections are strengthened, and others are removed. The neurons that are used frequently increase their connection strength, and the ones that are not used eventually die. This is known as **synaptic pruning**.

Keep in mind, too, that our degree of brain plasticity is believed to be influenced by our genetic makeup – so it's the interaction between genetics and environment that shapes the brain's plasticity.

There are two main types of brain plasticity. **Structural plasticity*** refers to the brain's capacity to change its physical structure as a result of learning and experience. **Functional plasticity*** is the ability of the brain to transfer functions from an area of the brain that is damaged to other, healthy brain areas. It's functional plasticity that has become one of the most groundbreaking areas of brain research in recent years. But the story gets even better than that.

The process by which new neurons are formed in the brain is called **neurogenesis**. This process is crucial when an embryo is developing, but it's now known that it continues in certain brain regions throughout our life span.

The diversity of neurons in the brain comes from regulated neurogenesis during the development of the embryo, where stem cells (special human cells that have the ability to develop into many different cell types) differentiate and become any one of a number of specialised cell types.

Adult neurogenesis was first recognised in the 1960s, but it wasn't until the 1990s that the scientific community accepted that neurogenesis in adult animals could play a substantial role in brain function. Taking up this idea, Professors Perry Bartlett and Linda Richards from the Queensland Brain Institute discovered in 1992 that the adult mouse brain contains neural stem cells. Since then, it's been found that neurogenesis occurs in the **hippocampus***and **amygdala*** regions of the adult human brain. This is an incredibly exciting discovery and the researchers believe this concept could hold the key to treating age-associated cognitive decline, neurodegenerative diseases including dementias, and other mental illnesses in the future.

The Ghost in My Brain: Clark's story

Clark Elliott is a professor at DePaul University and an expert in artificial intelligence and cognitive science. He sustained a traumatic concussion injury following a car crash and told me that afterwards he felt like a zombie. He still looked like himself on the *outside*, but *inside* it was as if his real self had left his brain and had become a ghost outside his body. He was afraid he would never see his original self again.

Disconcerting things happened to him too. At one stage he couldn't seem to get out of his chair, no matter how hard he tried. He unwittingly put his shoes on the wrong feet and walked around in them all day. He couldn't understand jokes, and he had to learn to sleep with his eyes open because closing them would cause dizziness and nausea. Being a cognitive scientist, he wanted to understand what was happening to him – so in the years after his injury, he took over a thousand pages of notes detailing how he felt his brain was responding to the injury.

Finally, he realised he could no longer read or comprehend passages he had just written. Doctors couldn't help him, telling him that he would never get better and he would simply have to learn to live with his symptoms.

Eight years after the accident, he could no longer cope. He was on the verge of losing his home, his job and the custody of his children.

As a last-ditch effort, inspired by Norman Doidge's first book *The Brain That Changes Itself*, Clark began to research the potential of the brain's neuroplasticity to heal his concussion. He stumbled across two brilliant researchers working at the cutting edge of brain plasticity: a cognitive restructuring specialist, Dr Donalee Markus, the founder of Designs for Strong Minds, and an optometrist, **Deborah Zelinsky** – the founder of the Mind-Eye Institute. Both practices are in suburbs of Chicago

In her practice, Zelinsky uses a neuroplastic technique she patented called the Z-Bell Test. She contributed to Clark's recovery by prescribing a special pair of glasses (known as Mind-Eye Brainwear) that was *not* designed to sharpen his central vision. Instead, the lenses were to activate his peripheral vision and match the way he perceived visual signals with how he perceived auditory signals. She told me she's helped many people who have a range of brain conditions, including ADHD, **ASD*** and learning difficulties by prescribing glasses designed for internal comfort rather than external eyesight and combining them with an individualised program of learning activities to rebuild dysfunctional visual skills.

As Clark explains, 'using these special brain glasses that Deborah Zelinsky prescribed for me, she found healthy pathways in my brain and she started routing the input signals from my retinas through these parts of this healthy tissue'. He adds that 'we can think of this as carving a dirt road through my brain, but through that healthy tissue'.

In the first few weeks of using these glasses, Clark felt like an eight-month-old child, rediscovering the world. He could feel new pathways again, but couldn't quite make sense of them, because his brain wasn't used to processing these signals through the healthy tissue. But it felt right, and he grew more used to it every day.

Clark also began treatment with Dr Donalee Markus, who's worked for over 30 years in the area of cognitive restructuring. She helped Clark reinforce the new brain pathways that had been created with the therapeutic glasses, by getting him to do complex brain puzzles, connecting the dots and rediscovering relationships between shapes, to turn the 'dirt roads' through his brain into metaphorical 'super-highways' that once again allowed him to be a professor of artificial intelligence.

After many months of these brain plasticity treatments, he noticed that the ghost lurking outside his body was getting closer and closer, 'and by the second week after its onset it was about six or seven feet away. By the third week it was just a few feet off my shoulder. I could never quite see it, but I could feel it. It was about as tall as I was, and shaped like a human. And in this moment I'll never forget, after teaching class, I was about to put the key in my door and I thought, "Oh, I get it, that's me coming home."

'I went into my office and put my head down on the desk and I just wept tears of joy that I could have this great gift again. And by the next day, Lynne, the ghost had entered me and I was once again fully human.'

Later, Clark used the notes he'd taken to write a book about his experience. And what else could he call it but *The Ghost in My Brain*?

Learning to compensate: Abbie's story

When **Abbie Kinniburgh** was two years old, her parents discovered that she had a rare brain disorder. She was diagnosed with partial **agenesis** of the **corpus callosum.***

The corpus callosum is a thick bundle of nerve fibres connecting the two hemispheres of the brain, which allows information to be transferred from one side to the other. It's important to any brain functions that need to be integrated between the two sides of the body, and for other functions, like reading and language, where one side of the brain needs to dominate and the other to be inhibited.

Corpus callosum agenesis is a developmental disorder in which the corpus callosum is missing. This usually happens to the foetus at around 12 to 16 weeks. Around 1 in 4000 people are affected. Some of them may be born blind or deaf, or may never learn to walk or talk – but others can function quite well in their daily lives.

Throughout her childhood, Abbie didn't notice any problems within herself, but her mother, **Maree Maxfield**, recalls being concerned.

As Abbie approached adolescence, she began noticing herself that she was different from others her age. 'I started to get very anxious about everything, just social situations and being a bit behind my peers in what I was doing. I went from a happy little kid that didn't have much worry to worrying about just about everything and not being very happy.'

As a teenager, she would still play with toys generally favoured by much younger children, and she had a need to be very organised and fastidious. Her peers began to exclude her, and the few friends she made were quite a bit younger than she was. She was a very slow writer and took a long time to process things, having to read passages over and over to understand them.

The doctors told her parents that she would probably never learn to read and write and would have lifelong learning difficulties. Fortunately, her mum and dad were both primary school teachers and kept encouraging her to aim high. 'My parents just kept letting me do things as I was growing up ... They thought, "Well, we are going to just keep aiming at the highest and just keep pushing a little bit further each time, and it doesn't matter if it takes a bit longer." So they never had the attitude of, "Oh well, she won't be able to do that so we're not going to try."'

Abbie said this was very helpful for her, so she didn't really notice how much the disorder affected her through primary school. With the continuous support of her parents and her own impressive inner strength, Abbie persevered through high school, and went on to study social work at university.

Now in her early 30s, Abbie still lives with a high level of **anxiety*** and a great deal of social difficulty. When I met her, she told me the most challenging thing for her was that people assumed that she *didn't have* any challenges: 'I often get told, "Well, you look fine, you act fine and you fit into the norms, so you mustn't have any challenges." So that's often frustrating, because I have a lot of challenges that people don't see. And the reason that I am so capable is because I've had 25 years to learn to fit in and cover it up, I guess, and I have been encouraged to do things, so I'm a lot better off than I could be.'

However, she's now active in talking about her experience and spreading her knowledge so that she can help and support others with this disorder, through the grassroots organisation the Australian Disorders of the Corpus Callosum (ausDoCC). Describing herself and other sufferers, she says, 'I think in a lot of ways we are very resilient and able to cope with adversity because we deal with it every day, almost.'

Abbie's mother, Maree, is currently a PhD candidate, researching personal quality of life among adults with corpus callosum disorders.

Corpus callosum disorders and neuroplasticity
Honorary Professor Linda Richards of the Queensland Brain Institute specialises in corpus callosum study, and is grateful to Abbie for her dedicated participation in this research. Richards is interested in the way some people with a corpus callosum

disorder – like Abbie – can cope quite well and lead full and productive lives in spite of the challenges they face. This has got Richards and her team intrigued about how the brains of these people compensate for their significant loss of brain connection.

Research has shown that the brain, when affected by a damaged or missing corpus callosum, creates plastic processes to compensate for the absence of the normal connection between the two brain hemispheres. Richards says some studies show that the fibres that would normally cross the corpus callosum actually find a new pathway in order to connect the hemispheres. This is why some people are able to function without a corpus callosum.

According to Richards, it's an example of a different type of plasticity from what is usually understood. Normally plasticity is thought about at the level of connections between individual neurons – but this type is known as **axonal plasticity**. This is the growth of completely new fibres across an entirely different tract of the brain.

As she explains, 'if we could understand how those plastic processes occur during brain development, to enable the brain to wire itself such that it can function in an optimal way, we may be able to repair the brain'.

For instance, 'one of the processes might be the kind of stimulation that a young child receives as it is developing. And so some work from my laboratory actually has shown that stimulation, sensory input, from both sides of the body is very important for wiring the corpus callosum … And so if we can understand those processes it might be possible to help make sure that that kind of stimulation is there as a child is developing so that we can maximise their potential later in life.'

Abbie's mum, Maree, says of Linda: 'She has been our guiding light, I suppose you'd say, so we are very indebted to her, and we hope that we can continue to work with her for a long time into the future.'

A word of caution

The science of neuroplasticity has opened the way for seemingly limitless hope of treating, healing and preventing distressing brain conditions. Brain plasticity does have its limitations, though. The brain is not infinitely malleable.

As I've discussed, certain areas of the brain are thought to be responsible for certain actions or functions such as movement, speech or cognition. If one of those parts is damaged, recovery may be possible, but other areas of the brain are not necessarily capable of taking over the functions of the damaged part.

However, as scientific knowledge increases, there is much excitement about the potential to harness the plasticity of the brain for beneficial outcomes. Scientific rigour is key in this field, and the growing movement to combine many diverse areas of expertise in collaborative, multidisciplinary teams will doubtless be invaluable.

2

THE MYSTERY OF CONSCIOUSNESS

Conversations with:

Martin Pistorius, locked-in syndrome sufferer

Joanna Pistorius, Martin's wife

Dr Michele Veldsman, postdoctoral research scientist in cognitive neurology, Oxford University; Director of Neuroscience (R&D), Cambridge Cognition, Cambridge

Bruce Goldstein, Associate Professor Emeritus of Psychology, University of Pittsburgh

Adrian Owen, Professor of Cognitive Neuroscience and Imaging, University of Western Ontario

Locked In: Martin's story

What exactly *is* consciousness, and how do we recognise it? What if we *fail* to recognise it?

For **Martin Pistorius**, these are far from hypothetical questions.

Martin was born in Johannesburg in 1975. At the age of 12, he developed a mysterious neurological illness. He fell into a coma aged 14, and was unable to move or communicate.

A couple of years later, he began to wake up – but nobody knew.

Martin will never forget the moment he became conscious, around the age of 16. He felt like he'd woken up as a ghost but didn't know he'd died. He couldn't understand why people were looking through him and around him. No matter how much he tried to beg, scream and shout, no one noticed. He had no control over any part of his body and was unable to communicate in any way.

Awareness began to come back to him gradually – until it dawned on him, to his horror, that he was probably going to be trapped inside his own body for the rest of his life.

During the years that followed, he could hear and observe things that perhaps he shouldn't have been privy to. They were sometimes funny, and sometimes very disturbing.

His mum had developed mental health problems and become suicidal. One night Martin heard her say that it was inevitable he would die. Of course, this was very difficult to hear, but he doesn't feel angry or resentful towards her – in fact, he feels enormous compassion. It was such a hard time for both his parents.

During these years Martin felt like he'd been stereotyped as an imbecile. He hated living in a care home. He's quick to point out that he did have some really kind carers over the years – but some were terrible. There was a lot of mistreatment, and he was even sexually abused.

But Martin was a fighter, and somehow he found an inner strength. He developed many strategies to help him cope with his darkest thoughts. He'd escape into his imagination – for example, fantasising about being very small and climbing into a spaceship and flying away or imagining that his wheelchair had magically transformed into a flying vehicle with rockets and missiles à la James Bond. In later years he spent a lot of time daydreaming about cricket. Go figure!

At last, after almost a decade in this isolating state, Martin reached a turning point. One of his carers, Virna van der Walt, became the catalyst that changed everything. She noticed that Martin could use very small eye movements to respond to things she said.

Around the age of 25, Martin was taken for an assessment and demonstrated that he had the potential to communicate. Virna's faith in Martin opened up a whole new world of computer-assisted communication for him, and finally allowed him to tell his story to the world.

Locked-in syndrome

Martin Pistorius has **locked-in syndrome.*** Neuroscientist **Dr Michele Veldsman**, now at Oxford University, explained to me that it's a condition that mimics a vegetative state, but in fact the patient is completely conscious, so it can be a devastating affliction. Frequently all they can move is their eyes. Locked-in syndrome is quite rare, and scientists still don't have good figures on how many people experience it.

Veldsman told me the syndrome typically occurs after damage to an area called the **ventral pons**, part of the **pons**, which is a major division of the **brainstem**. It can be the result of tumours, traumatic brain injury or stroke. Martin's case is highly unusual, in that no cause for his locked-in condition has been found.

Veldsman described to me the computer-based communication system that changed Martin's world. For patients like Martin, 'brain–computer interfaces really are vital. At the moment these are done with **electroencephalography**, so **EEG**, putting the electrodes on the scalp. And the kind of paradigm is to have the patient think of the letter that they want to spell out, the word that they are trying to spell out, and then they are shown a series of letters, and there will be a signal that shows familiarity, because every other letter will be a surprise, and when the letter is shown that they are thinking of, this will register a kind of signal of familiarity. And this is used then in sophisticated computer algorithms similar to predictive text messaging to be able to communicate and build out words and sentences.'

My conversation with Martin was an extraordinary experience for me. We spoke via Skype between the United Kingdom and Sydney. Because he was using this computer-generated voice technology, his responses to my questions were quite delayed. It reminded me of conversations I'd heard with the late physicist

Stephen Hawking, who had a form of motor neuron disease. But despite the stilted feel of our chat, it was authentic and thoroughly engaging.

This sophisticated technology also opened the way for Martin to meet the love of his life.

Joanna came to Martin's residential care home to visit a friend. Eventually she and Martin were introduced and there was an instant connection. When she went home, they began to talk to each other every day over Skype, using computer-generated voice technology – he in South Africa, she in the United Kingdom.

Martin realised he had met someone who could teach him the importance of having dreams for his life. The relationship blossomed, he migrated to the United Kingdom to be with her, and in 2009 they married. Martin says that one of the proudest days of his life was when he graduated with a computer science degree in 2013. He now runs a web design business from his home in Essex.

And in 2018 – despite having been told that they could never have children – to their great joy their son, Sebastian, was born, whom Martin describes as a 'pure, jolly and loving little soul'. He explained to me that while it was exciting to be a father, it was a bit daunting, especially around communication and being able to support a family. 'We have used baby signing basically from birth, which worked well. The past few years have not been easy, but it's an amazing experience. I also feel that because of what I have been through I have some insight into what it is like to try to make sense of the world and learn to communicate. Sebastian and I are really close.'

Martin went on to say that 2018 and the year leading up to then was quite a year for him. 'I competed in wheelchair racing events in Switzerland and to my surprise set a European record for the 1500 metres in my class. I also received an honorary doctorate!'

For her part, Joanna says, 'I've got a lot of respect for Martin. I just think he's such an amazing person. He's so determined. But also I think he's such an honest and sincere person that whatever he faces he's honest about it and sees what is the best he can do about it.'

I asked Joanna what insights she'd gained from Martin's experience.

'Anything is possible,' Joanna told me. 'And not to focus so much on the barriers; focus on what somebody can do and build on that, and keep on dreaming, never give up, even if it's small things, dream about it and start focusing on how you can achieve those things, because one dream leads to another.'

What is consciousness?

I'm looking out at a beautiful seascape. The sky is a dazzling cerulean blue; the aquamarine ocean is flecked with bright white waves and glistening reflections from the sun. Nearby there are bottlebrush trees, covered in rich red flowers.

If you were standing next to me, could I be sure that you were experiencing the same thing as I am?

The answer is clearly no. Even though we may *think* we share the same experiences, our individual minds are unique.

'What is consciousness?' is probably one of the hardest questions we could ask. It's a mystery that has baffled, intrigued and challenged scientists and philosophers for millennia, and is at the heart of my deep fascination with the human mind, and why we behave the way we do.

What most of us would agree on is that consciousness is our own distinctive awareness of ourselves and what's around us. It includes our thoughts, emotions, sensations, memories, feelings and perceptions, and these shift and change from one moment to the next. We have a long way to go in understanding its exact nature, but the knowledge that's come with advances in neuroscience over recent years is truly tantalising. The study of consciousness is currently one of the most exciting and dynamic fields in science.

A leading thinker in this field is Australian philosopher and cognitive scientist David Chalmers, currently Professor of Philosophy and Neural Science and Co-director of the Center for Mind, Brain, and Consciousness at New York University. In his 1995 paper 'Facing up to the problem of consciousness', he introduced what he called **the hard problem** of consciousness: the problem of explaining how all the complicated physical and chemical processes that take place in our brain produce our unique feelings and perceptions, our desires, hopes, fears, pains and pleasures. Why doesn't all that brain activity just go on in the dark, as if we were robots?

There are also what he refers to as **the easy problems**. These are about finding out how the physical processes in the brain give us the ability to think, take in and sift through information, to control our behaviour and perform other cognitive functions. They are considered easier problems because it is possible to study them objectively, using standard scientific methods. Even though we don't have anywhere near a complete explanation of these phenomena, scientists at least have some idea about how they might go about explaining them.

Chalmers points out, though, that 'easy' is a relative term. He admits that whoever solves these 'easy problems' of consciousness will probably be up for a Nobel Prize!

Brain vs mind

One of the oldest dilemmas in the study of consciousness is the **mind–body problem**: what is the relationship between the physical and the mental – between the brain and the mind?

Bruce Goldstein has devoted his life to studying the senses and cognition. 'The mind and the brain are intimately connected,' he told me, 'because the brain creates the mind. However, there's a big difference between them. It has often been said that the brain is something that's material. You can take a picture of the brain, it can be photographed, but you can't take a picture of the mind. The mind is invisible.

'And so that has led to a lot of arguments. There are some people who say that they are two totally different things, but most respectable neuropsychologists [who study the effects of brain injury and illness on cognition and behaviour] say that the mind depends on the brain.'

How, then, does consciousness fit into the picture? The sensation of consciousness is the only thing that tells us we exist, but where does that sensation come from? Is it from outside our physical body, or is there a centre of consciousness in our brain?

There's a vast amount of scientific research going on across the world that is trying to solve all of these 'hard' and 'easy' problems.

Detecting consciousness with neuroscience

Martin Pistorius found amazing resilience within himself to face the nightmare he lived through, but imagine what a difference it could make to *others* assumed to be in a vegetative state if their consciousness could be detected earlier.

This is what British neuroscientist **Professor Adrian Owen**, now based in Canada, has been investigating throughout his career. In 2006, he published a paper called 'Detecting awareness in the vegetative state', revealing his finding that some patients who appeared to be in an ongoing vegetative state were in fact fully aware, and able to communicate using brain imaging technology.

'It's a story about serendipity, really,' Adrian explained to me. 'I trained in neuroimaging and neuropsychology. And then in 1997 I was working at the University of Cambridge and I was introduced to a patient whose name is Kate Bainbridge, who was in a vegetative state.' However, 'back then very few vegetative patients had gone into brain scanners because people just assumed you would see nothing, there would be nothing going on in their brain.

'And Kate was an amazing case, because we scanned her, and while she was in the PET [positron emission tomography] scanner we showed her pictures of faces of her friends and relatives, and the part of her brain that we know is involved in processing faces lit up just as it would in a healthy participant. So there was a response in her brain to familiar faces. And at the time this was amazing, it was completely unexpected and very, very exciting, and it was the first time I think that imaging had been used to show that some vegetative patients have some residual cognitive activity in their brains, and that really started the whole thing rolling.'

Later Owen and his team were able to establish that patients like Kate could hear and comprehend speech. The areas of the brain responsible for this will receive more blood flow and light up when patients are in a brain scanner. So, neuroscientists are able to map the patient's brain according to what areas are being activated in response to different types of external stimulation.

However, while Owen's team realised that their patients' brains could respond to speech, they still didn't know whether the patients were actually *conscious*. That was when they conducted a classic experiment.

In TV medical dramas, the doctor will often say to a seemingly unconscious patient, 'Squeeze my hand if you can hear me', and the patient will respond with a squeeze. Obviously this wouldn't work with people who were paralysed, so instead, Owen's team wondered if they could get a patient to *activate their own brain* when asked to do so.

This was how they did it. They got patients to imagine playing a game of tennis.

There's a part of the brain known as the **premotor cortex** that sends messages to the **motor cortex**, telling the motor cortex to move a part of the body. Even if we don't actually move that body part, but just *imagine* doing so, our premotor cortex will still be activated.

So, when paralysed patients were put into a brain scanner and asked to imagine playing tennis, their premotor cortex lit up. Thirty seconds later, they were asked to stop playing tennis and the activation disappeared. It was a bit like saying, 'Squeeze my hand if you can hear me', but really it was *'Activate your premotor cortex* if you can hear me.'

Owen's team continued their research and developed other ways of helping vegetative patients communicate, by imagining different things to change their brain activation state – even telling them jokes.

Research now shows that around 20 per cent of patients thought to be in a vegetative state are in fact locked in, or minimally conscious but unable to respond. This raises important ethical questions about the need for better diagnosis and subsequent treatment of these patients.

'The question that is closest to my heart, though, that I think is raised by this,' says Owen, 'is whether we can as often as possible give some autonomy back to the patient. The really heartbreaking thing about this group of patients is that decisions about their care and decisions about their future and often decisions about whether they live or die have to be made by close family members, and that is an awful situation.' Owen is confident this will start to change very soon.

The case of Martin Pistorius is a compelling example of how research like this can make a real difference to the quality of life of people faced with the devastating impacts of brain conditions affecting their consciousness.

Martin has reflected on the insights he has gained from his harrowing experience. He says: 'I think that there is always hope, no matter how small. Treat people how you would want to be treated, with kindness, dignity, compassion and respect, whether you think they understand or not. Never underestimate the power of the mind, the importance of love and faith, and never stop dreaming.'

3

MEMORY

Conversations with:

Suzanne Corkin (1937–2016), Emeritus Professor of Neuroscience,
Massachusetts Institute of Technology

Jenni Ogden, novelist and retired neuropsychologist

Elaine Reese, Professor of Psychology, University of Otago,
New Zealand

Suparna Rajaram, Distinguished Professor of Psychology,
Stony Brook University, SUNY

Bruce Goldstein, Adjunct Professor of Psychology,
University of Arizona

Professor Karen Adams, Director, Gukwonderuk Indigenous Health
Unit, Monash University, Melbourne

Lynne Kelly, science writer and Adjunct Research Fellow,
La Trobe University, Melbourne

Rebecca Sharrock, hyperthymesiac

Gail Robinson, Professor of Psychology,
University of Queensland

The most valuable brain in the world: The story of 'HM'

Memory is as elusive and difficult to define as consciousness. Scientists have wrestled for centuries with what it is, how it works and how reliable it is. The way we conceptualise memory has changed dramatically over that time. We now understand that it involves a very complex web of brain cells and systems that records information that comes into the brain and stores it for later retrieval. (Memory is such an important subject that I'll refer to it again and again in this book.)

Even as late as the mid-20th century, scientific knowledge of memory was quite limited. But one famous brain would soon change all that.

'HM' was born in 1926 in Connecticut, and developed severe epilepsy at a young age. In 1953, he was referred to Dr William Beecher Scoville, a neurosurgeon at Hartford Hospital.

Scoville's diagnosis was that HM's epilepsy was caused by his left and right **medial temporal lobes**, located behind his ears. The medial temporal lobes of the brain include the **hippocampus***, **amygdala*** and parahippocampal regions, which are crucial for episodic memory, or memory for everyday events, as well as spatial memory. So two holes were drilled into HM's skull, and parts of his medial temporal lobes – the front part of his hippocampus and most of his amygdala – were sucked out.

We now know that the hippocampus and the amygdala are two of the places where long-term, explicit (conscious) memories are stored.

HM's fits did improve slightly, but the unexpected and devastating consequence of the operation was that from that day on he lost the ability to lay down new long-term memories. Consequently, he lived permanently in the present tense, only ever able to remember the last 30 seconds. His world was turned upside down, yet little did he know that this was the beginning of his unprecedented contribution to our scientific understanding of memory.

When his case was first reported, he was referred to as 'HM'. Thousands of psychology students would come to know him by this name.

It was only after his death in 2008 that his identity was finally revealed as Henry Molaison.

The study of permanent present tense

The late neuroscientist **Suzanne Corkin** worked intensively with Henry for over 50 years at the Massachusetts Institute of Technology (MIT), yet he never knew who she was. She told me Henry's fascinating story, which she wrote about in her book *Permanent Present Tense: The Man with No Memory, and What He Taught the World.*

Corkin explained that what made his case invaluable was that his amnesia was caused by precise surgical intervention, not by brain disease. This meant that all his other intellectual functions were intact, his language and perceptual skills were normal, and his IQ was consistently above average. He had no psychiatric symptoms; he just had an exceptionally bad memory.

Immediate memory is something that never leaves your consciousness – it's right now. Henry's immediate memory was preserved, so he could repeat back to Corkin a string of numbers straight away because they were still in his consciousness. However, once she left the room or he was distracted for a minute he could not remember them.

Henry's memory loss had some interesting side-effects. For one thing, his body image was outdated: he described himself as thin, but he was now heavy, and he was unaware that his hair had turned grey. He never knew when he was hungry or thirsty, or whether the next meal was breakfast, lunch or dinner. His pain threshold was very high, possibly because of the amygdala lesion; the amygdala plays an important role in our response to pain. So, for example, Henry had very bad haemorrhoids for many years, but he never complained about the pain.

Corkin described Henry as a 'gentle soul', an always friendly research participant who was keenly motivated to help others.

New Zealand neuropsychologist **Jenni Ogden**, who worked with Henry during a year-long MIT fellowship, agreed: 'Of course you'd give him these boring experiments and the same things over and over again, but he was never bored because he'd never seen them before. That was one of the wonders about HM, he was such a joy to work with. You know, you couldn't help but be fond of him.'

Corkin told me Henry's case launched the modern era of memory research by highlighting three major principles. Firstly, Henry was living proof that you can be an intelligent person and still have a dreadful memory. Secondly, Henry proved that not all kinds of learning and memory are impaired in amnesia.

Thirdly, before Henry it wasn't understood that the ability to form new memories is localised to specific parts of the brain.

The finding by neuropsychologist Brenda Milner that different memory circuits are confined to different brain areas was ground-breaking. In 1962, Milner showed that Henry could learn a motor skill that he practised over three days, even though he never remembered the test apparatus, how well he had performed or Milner herself. This finding opened the way for extensive research into the idea of multiple memory circuits.

Corkin's observations and research with Henry revealed further interesting insights into the role of the hippocampus. This brain structure is very important for encoding new information, consolidating it, then helping store it. Some memories are eventually transferred as general knowledge from the hippocampus to the **neocortex.** (This is the largest part of the **cerebral cortex,*** the wrinkly sheet of grey matter on the outside of the brain.)

Henry in fact retained all his general knowledge from before his operation: he knew about World War II and the Great Depression, because his awareness of these events was still stored in his neocortex. Corkin believed that was also why his vocabulary was so good. However, he did not have a *detailed* memory of world events before the operation, because the hippocampus is essential for retrieving explicit memories that happened at a particular time and place. Corkin and her colleagues worked hard at trying to help Henry remember anything specific about his parents, a birthday, an illness – something that happened once and only once – but he couldn't do it. He was just left with the gist of his pre-operative life, like different addresses where he'd lived, his hobbies or the names of his neighbours.

There were also some things about Henry's memory that Corkin and colleagues *didn't* expect. Because he spent a lot of time watching television and reading magazines, he could give accurate distinguishing information about a small number of people who had become famous *since* his operation.

In 2004, researchers Gail O'Kane and Elizabeth Kensinger conducted an experiment in which they showed Henry pairs of names – one of them famous, and one picked from the Boston phone book. Some of the names were chosen from before his 1953 operation, and others from post-1953. He was asked to say which people were famous.

Not surprisingly, 92 per cent of the time he knew which of the two names was famous from *before* his operation, but unexpectedly, he was correct 88 per cent of the time about the *post*-operative names.

Then O'Kane and Kensinger asked him *why* these people were famous – and for about eight people he could give some facts. These were people who became famous after his operation, in the 1960s, 1970s and 1980s. For example, he knew that Julie Andrews was famous for singing on Broadway, that JFK was a US president who was assassinated, that Woody Allen was a comedian in movies and that Liza Minnelli was a dancer and a movie star.

Corkin explained to me that these memories were retained by Henry because they had an *emotional* component for him – they particularly appealed to him or entertained him. This has cast some light on the connection between emotions and memory.

Corkin told me research shows that our memory of emotional information is generally much better than our memory of neutral information. For example, most people in the Western world have vivid memories of what they were doing when they learnt about the 11 September 2001 United States terrorist attacks. In contrast, for many people in the United States, the Super Bowl football final is a big deal, but people don't tend to remember the details unless it was *their* team that won or lost.

HM's sense of self

Having worked so closely with Henry over many years, did Corkin agree that because he lacked the ability to remember, he also lacked self-identity?

Henry *did* have a sense of self, she felt, but it was less complete than most people's. Mostly our notion of self is a composite of past memories, present thoughts and future plans, but for Henry, access to all three was patchy.

He had no personal autobiographical memories from after his operation. For a long time after his parents died, he wasn't really sure whether they were alive or not, so he carried a little note around in his pocket. He did, however, have fragments of information about his own situation. For example, he was aware that he had had an operation on his brain and knew that something had gone amiss, which meant that he had a bad memory.

When it came to looking into the future, Henry was at a loss. When Corkin asked him what he would do tomorrow he would say, 'Whatever is beneficial' – full stop. He could not chase future dreams, because he didn't have any.

HM's legacy

Henry passed away in 2008. It was a very emotional time for Corkin, as she had become very close to him. But she and her colleagues had already realised that Henry's brain was perhaps the most valuable in the world. So an extensive autopsy, which had been planned well in advance, was carried out immediately.

Henry's contribution to our understanding of memory has inspired many students to pursue careers in neuroscience. Corkin told me: 'I would thank him of course for what he was doing and I would tell him how important it was, and I would say, "You know, Henry, you're famous. People all over the world know who you are and they know about your wonderful contributions to science and how things are going to be better for other patients because of you."

'And he would just sort of smile sheepishly, sort of took it in his stride, and I'm sure it warmed his heart. But naturally the minute he was distracted he forgot it.'

So sadly, the world will remember HM and his contribution to science much better than he ever did.

Enriching children's early memories

HM lacked a clear sense of self, because he had no autobiographical memories – and **Professor Elaine Reese** of the University of Otago, New Zealand, is fascinated by how our autobiographical memory makes us unique. This led to her work on a longitudinal study called 'Origins of memory', which has followed over 150 New Zealand children from the time they were toddlers up until they were 21 years old. Researchers are finding that children's earliest memories are influenced by their sense of self (which develops between about 18 and 24 months), by the way their parents talk to them about their memories, and by their own language capabilities.

One of the most important things parents can do to reinforce their children's memories, according to Reese, is to use open-ended questions. 'So for instance, say you took your toddler to the zoo and the next day you might ask your toddler what animals did we see at the zoo? And a lot of toddlers won't say much in response but then if the parent follows up with "What animal did we see that gave a big roar?" … So you're giving new information … Parents who ask more of those kinds of questions, their toddlers grow up to be pre-schoolers who have richer memories.'

Reese and her colleagues think this gives children a structure for keeping the memory alive. 'So we think that they've already laid down the memory, but if parents talk about it after it's happened then what it's doing is helping the toddler to put that memory into their own words. And we know that just the act of putting it into language is helping to strengthen that memory soon after it happens, it's probably helping to consolidate that memory and then that memory will last for longer.'

The researchers are finding that adolescents whose mothers talked to them in this 'elaborative' way in early childhood have memories that date back to a younger age than adolescents whose mothers *didn't* use this style of reminiscing. They also have more *coherent* memories – they are able to talk about more difficult events in their lives in a more structured and meaningful way.

These young adults with a more coherent life story have better wellbeing and lower rates of **depression*** and **anxiety.***

Evidence for the long-term benefits of reminiscing in an elaborative way with toddlers has continued to grow. Reese and her colleagues now know from their longitudinal intervention study called 'Growing memories' that reminiscing has enduring effects for teens' and young adults' wellbeing. They've found that in 21-year-olds it lowered their risk of depression, increased their self-esteem and improved their understanding of difficult life events.

Telling family stories

In Reese's book *Tell Me a Story: Sharing Stories to Enrich Your Child's World*, she examines the role that telling family stories plays in the development of children's memories.

Family stories are different from autobiographical memories, since they're often about events that the child didn't personally experience. 'So one popular one,' says Reese, 'is telling the story of the child's birth, and that can be a really positive family story, talking about how the child came into this world, into this family, and all the people who were so glad to welcome that child.

'As children get older parents can start to tell stories about when they were young, and you'll know it's working if they ask to hear the same family story again. A story from the parent's own childhood may also help a child cope with a particular problem that they're having in their lives right now. Many of the stories can be enjoyable and funny, and they can be quite bonding for families.'

Further research by Claire Mitchell, Sean Marshall and Elaine Reese from the 'Growing memories' study showed that coaching mothers to converse with their children about everyday memories in a detailed and affirmative way, and speaking about life's turning points (such as parental divorce or cyberbullying), enhanced their wellbeing and lowered levels of anxiety and depression through adolescence and into young adulthood.

So, Reese's message is: talk to your kids in an 'elaborative' way from the word go, and don't hold back with the family stories. For young children, storytelling is important for their language development and their memory, and as they get older, it's very important for their wellbeing and self-esteem.

Improving memory

Should you cram before an exam? Does repeating information to yourself help you retain it?

These are the kinds of questions that intrigue **Suparna Rajaram**, Distinguished Professor of Psychology at Stony Brook University, SUNY. In her extensive memory research, she has drawn on people with typical memory abilities, as well as those who have a memory deficit.

Cramming and rehearsing

Rajaram has looked at the effectiveness of cramming before an exam, or repeated study in a short space of time. Cramming often gets you through the immediate test but you're unlikely to remember that information into the future. Rajaram and her team wanted to understand why knowledge from repeatedly studying dissipates over time.

They found that repeated study *does* improve retention in the short term, but a lot of that information retained is not organised tightly or efficiently. Instead, repeatedly testing ourselves is the key.

This kind of retrieval allows us to align what we have just learnt more efficiently with our pre-existing knowledge.

This has implications for the way children learn at school. While Rajaram suspects that tests are not popular, she thinks they're a good idea. Repeated study can fool us into thinking that a larger amount of information has been retained. Repeated *tests* might focus on a smaller amount, but they help us get down to the important pieces that we can hold on to. But Rajaram adds that it's important for students who are taking repeated tests and giving incorrect answers to be corrected, or they will retain the *wrong* information.

The ageing brain

Rajaram believes that repeating information could also help older people's memory. Repeatedly going over something in our mind is like repeated testing, because we are generating the information ourselves. If somebody *tells* us what happened, it is less effective than thinking about it ourselves and narrating it repeatedly to others.

Our memory and learning ability change significantly throughout our life span. In adolescence, our working memory and planning and decision-making skills increase significantly due to the development of the **prefrontal cortex***, the part of the brain that controls these functions. The prefrontal cortex continues to develop till the age of about 25, so our ability to remember new information peaks in our 20s, then noticeably decreases in our 50s and 60s.

It's believed that this decline is caused by the deterioration of the **hippocampus;*** by a decrease in the hormones and proteins that repair and form new brain cells; and by decreased blood flow to the brain. Even a healthy ageing brain starts to exhibit some memory deficits, mainly to do with episodic (day-by-day) memories.

But it's not all bad news: older people's ability to *organise* information remains fully intact. One thing we often don't bear in mind is that when we get older, we know more, which means we have so much more information to navigate to get to the right memory, or the right answer, or the right name. So we jump to the conclusion that our memory is no good any more.

And when it comes to the ageing brain and memory, Rajaram reinforces the advice we often hear: learning something new – another language, a musical instrument – and exposing yourself to new situations can be very helpful in protecting against the memory decline that's natural with ageing. (I'll have a lot more to say about ageing and memory decline when I talk about **dementia**,* in Chapters 11 and 20.)

Using the past to predict the future

Amazingly, the consensus among neuroscientists is that one of the important roles of memory is to help us predict the future. We recall that when we did a particular thing there was a good result, but when we did a different thing, the result wasn't so good. That information helps us behave more efficiently and successfully in the future.

Bruce Goldstein – whom I introduced in the previous chapter – told me about a neuroscientific experiment in which the researchers measured the brain activity of a person remembering a past event. Then they asked the person to think about a similar event that was happening, say, a week or two in the future. When they did that, brain activity appeared in the same areas. This highlights the very close link between memories from the past and what we predict is going to happen in the future.

One of the hottest topics in psychology today, Goldstein suggests, is a concept called **prediction error**. Let's say you're taking a walk through the bush, along a track you've often followed. You're familiar with many of the sights and sounds you encounter – but suddenly you hear a strange, unexpected sound coming from your right. You realise that the usual predictions you make as you walk along this track are wrong, and that you have to adapt your behaviour to counter that error.

Any time something happens that is not what we expect, the brain generates a prediction error, then we have to take action to correct it and adapt to the new reality. There are many mathematical models being used in brain science to try to explain exactly how this works.

Language is another area in which the brain makes predictions. Most of us would have observed that people who have lived together for a long time often seem to know what the other person is going to say before they say it. We make predictions when we read too. If, for example, you read 'John put on his swimmers and decided to' you'd predict that the rest of the sentence will have something to do with jumping into the water. But if the next words were something surprising like 'go to his job interview', this would cause a prediction error that you would have to adjust to.

Our brains are constantly making predictions like this as we go through our lives. And this predictive power is essential to our survival.

Indigenous memory code

Indigenous Australians have the longest continuous culture in the world. It may date back more than 60,000 years. Like oral people everywhere, Australian Aboriginal people use cues from the landscape to remember the past and pass on knowledge, cultural values and wisdom.

Information about thousands of species of plants and animals is intricately tied to the landscape and operates as a potent form of cultural memory that is passed from one generation to the next without ever being written down.

Karen Adams is a Wiradjuri woman and Director of the Gukwonderuk Indigenous Health Unit at Monash University in Melbourne. She explained to me that Aboriginal history is passed down through the intricate song, dance, art and stories of the Dreamtime, the time when ancestral spirits created the land and its creatures.

Within this history is the oral tradition of **songlines**, an ancient memory code used by indigenous cultures throughout the world. Within Australian Aboriginal culture, they relate to particular places and represent creation stories, contemporary stories and stories about social law. 'It's quite complex, but those land markers are very, very important, and hence the importance of land claims and acknowledgement of traditional owners,' Adams says.

She shared with me a story from her childhood that she's remembered vividly throughout her life. She explained that Wiradjuri means 'three rivers', so her people see themselves as river people. They tell a story that has different versions, about a duck and a water rat. In some versions the water rat steals the duck; another variation is that the water rat and the duck fall in love. The duck becomes pregnant and has babies, but her family ostracises her, because the babies are a mixture of duck and water rat. She leaves her babies in different places along the river. Those babies are platypuses.

Adams says it's a social law story about appropriate relationships, but also about the Wiradjuri people's relationship to those particular animals. Every time Adams sees platypuses she remembers those animals and the river. Stories like this establish a deep connection between Aboriginal people and their environment, and highlight the cultural importance of caring for country.

Learning memory codes

Science writer **Lynne Kelly** told me she is blessed with an extremely bad memory, and became fascinated by how Aboriginal people remember the thousands of plants and animals in their environment. She immersed herself in the memory methods used by ancient cultures to retain vast amounts of information about their local country. By collaborating with Australian Indigenous people, including her colleague Nungarrayi, a Warlpiri woman, she gained an insight into their encyclopaedic memory.

She describes songlines as navigational tracks that the Elders
and trained Indigenous people follow through the landscape.
They move from location to location, singing information and
performing rituals about each sacred site. Research has shown
that up to 70 per cent of the songs consist of knowledge about
animals, plants and seasonality: all the information you need to
know to survive in a particular environment.

In songlines, each location in the landscape acts as a mnemonic,
a memory aid for a particular part of the information system. The
knowledge is literally grounded in the landscape. It is supported
by different forms of portable art, such as message sticks and
tjuringa, sacred objects made of polished stone or wood. Songlines
use sequences of places, as well as these devices, to order the
memories, so that none of them are lost. Kelly says that Aboriginal
people don't need to be walking the songline to recall it, as the
memories are so strong.

I must say I've noticed that if I'm walking in a particular
environment while listening to a podcast, remembering parts of
what I've heard will also trigger associations with the landscape
I've just walked through. The system fits perfectly with the
way the human brain works: in 2014, John O'Keefe, May-Britt
Moser and Edvard I Moser won the Nobel Prize in Physiology
or Medicine for their research showing that the human brain
associates memories with particular places.

After studying the way memory is embedded in the landscape
of many cultures, Kelly drew on these techniques to develop
her *own* memory code, using the environment around her, to
remember swathes of information she was not otherwise able
to recall.

She started with remembering the countries of the world. She
associated 10 different locations in her office with the 10 most
populous countries. Then she went out into the garden, around
the house, and down the street and back. Each house and location
she passed represented a different country for her. Now, she says,
when she's watching the news, a country comes up and her brain

automatically goes to that place. She doesn't have to remember the countries in sequence, because they're fixed in sequence by the landscape, and she can add more and more information because there is already a structure in place.

She's also remembered history, starting 4 billion years ago, by walking around prehistory – which takes her about a kilometre.

To remember bird species, she's encoded a complete field guide to the 408 birds of Victoria by using a stick, something like the Aboriginal *tjuringa*. She creates stories around each of the birds and glues particular shells onto this stick to remind her of the stories. Kelly believes that these techniques also strengthen her memory for everyday things.

Does she ever feel overwhelmed, I asked her, with too much information stored in her head? She doesn't, she said, because they're images, not words.

'Orality is what people have when they don't have literacy,' she told me, 'and it's also incredibly powerful. I would love to see schools teaching a combination of the two.'

Recalling every day of your life: Rebecca's story

I'm sitting high on my father's shoulders. It's a bright sunny day, and we're peering across the crowds at the opening of the new local swimming pool. My dress, which my mother made, has small rosebuds embroidered on its bodice. Dad and I are both squinting and frowning, and I'm feeling grumpy.

I think I was around three when this happened. The memory is crystal-clear. But is it an *authentic* memory, or has it been aided by a family snapshot, or the story often told by my parents?

If you're anything like me, your earliest recollections will feel rather hazy. Recent research suggests that, on average, people's earliest memories are from when they were around two and a half years old. Little is known about why; perhaps the earlier memories are still there but inaccessible to us, possibly obscured when new brain cells are generated.

Rebecca Sharrock is different. One of her first memories happened on 23 December 1989, when she was just 12 days old. She was lying on the sheepskin cover of the driver's seat in her parents' four-wheel drive, and they were taking a picture of her. She was looking up at the steering wheel and the camera, wondering what they were.

As Rebecca tells me this, she becomes quite emotional. She remembers how, at 12 days old, every sight, sound and scent was a wonderful novelty.

She has some even earlier recollections than this. She doesn't remember leaving the womb, but she recalls being in a bed on a cotton blanket with glass walls all around her. She was looking at the ceiling and the heads of the people who were dressing her and putting blankets on her. She can't say for certain that that was the day she was born, but she guesses it was around that time.

Rebecca can't forget a single day of her life. She is one of approximately 60 known people in the world, and the only one in Australia, with this kind of exceptional memory. It's known as **hyperthymesia,** or **highly superior autobiographical memory (HSAM).*** Rebecca is the single participant in research on HSAM led by **Gail Robinson**, Professor of Psychology at the University of Queensland.

HSAM was first identified in 2006 at the University of California, Irvine, and Robinson is working with the researchers there to understand the underlying neural basis of the condition. She describes HSAM as a very rare and selective type of memory ability that allows someone to recall an exceptionally high number of autobiographical and episodic (day-to-day) memories, and frequently the dates when they occurred. Often people are unaware that they have this exceptional ability until they hear about it, perhaps on the radio or TV. Rebecca thought her memory was completely normal until she was in her early 20s, when she and her family saw a TV documentary describing HSAM. Rebecca also has **synaesthesia,*** a condition I'll examine in Chapter 8.

In my conversation with Rebecca, I picked a few random dates over the course of her lifetime and asked her what she remembers about them.

The first date I chose was 4 September 2006. She quickly replied that it was a Monday, and she and her mum were on their way to Rebecca's therapist when her stepdad called to tell them that the Australian conservationist Steve Irwin had died after being stung by a stingray. 'And I couldn't believe it,' says Rebecca.

The next date I gave her was 24 June 2010. She immediately said it was a Thursday and she didn't come across any news that day, but when reading the papers she noticed that an oil spill in America and volcanic eruptions in Iceland were still in the headlines from a few months earlier.

Rebecca knew that 22 January 2008 was a Tuesday, and she recalled feeling quite strange that day. All her younger siblings were getting ready to return to school, but this was the first year when she wasn't going with them, because she'd graduated at the end of the previous year.

Rebecca is also a big fan of the Harry Potter books, and I was told she remembered them word for word. To check this out, I randomly selected a few lines from a couple of different books in the series and asked her to recite the sentences that followed.

My first quote was: 'October arrived, spreading a damp chill over the grounds ...'

She responded with: '... over the grounds and into the castle. Madam Pomfrey, the nurse, was kept busy by a sudden spate of colds among the staff and students.' She added that this was from *Harry Potter and the Chamber of Secrets*, the second book in the series, and it was Chapter Eight, 'The Death Day Party'.

The next line I gave her was: 'Harry went down to breakfast the next morning to find the three Dursleys already ...'

She followed with: '... already sitting around the kitchen table. They were watching a brand-new television, a welcome-home-for-the-summer present for Dudley, who had been complaining about the long walk between the television and the fridge in the living room.' She told me it was from Chapter Two, 'Aunt Marge's Big Mistake', in *Harry Potter and the Prisoner of Azkaban*, the third book in the series.

I was astounded at how confident and accurate she was!

Robinson says there is some evidence of structural differences in the brains of people with HSAM. Limited studies using brain scans show possible differences in the **autobiographical memory network**, which includes the **temporal pole** and **lobes**, the **hippocampus**,* the **insula** and the **caudate putamen**. However, these findings are not conclusive and it's not known whether these potential differences are *the cause* of HSAM or *the result of activities that cause* HSAM.

Another question the researchers have considered is whether Rebecca's other cognitive abilities, such as other types of memory, intelligence, language, perception, complex thinking and problem-solving, are superior as well. So far, they have found that she performs in the average range on these other cognitive tests. So something exceptional is going on, but it's not explained by other cognitive abilities.

Gift or curse?

Rebecca's autobiographical memory is very impressive, but sometimes her recollections can trigger strong emotions. With a positive memory she relives happy times, but if her memories are from a time when she was depressed, she relives **depression**.*

Once she was walking down a pathway and saw a leaf on the ground that was at a similar angle to another leaf she'd seen when walking home years ago, after she'd been bullied at school. This triggered the feelings of hopelessness that she'd experienced back then.

Reliving physical pain is also a real challenge for her. She recalls grazing her knee when she was three years old, and how she cried because it burnt and stung so much. For a young child, an injury like this hurts much more intensely than it does for an adult.

Living with so much information and emotion in her head is often tiring and overwhelming, and she gets frequent headaches. She has to have the television on or music playing to fall asleep, otherwise she's kept awake in the dark and quiet by a cacophony of involuntary memories.

When Rebecca first found out she had HSAM, she thought she was broken. She felt there must be something wrong with her because she couldn't get over events from so many years ago. She was too embarrassed to explain why she got so upset when she relived the memory of having another child take her lollipop away from her as a seven-year-old.

More recently, after getting help from therapy and as she's progressing in her career, she sees her HSAM as a balance of pros and cons.

When I spoke to her she was in the process of writing her autobiography. Unsurprisingly, she was finding the writing and editing process overwhelming, because it forced her to relive all the emotions associated with her memories. However, she loved writing about her happy years, about holidays and good times, because she would relive those positive memories in all their detail.

HSAM research and dementia

Professor Robinson believes the study of this amazing memory ability may hold some important insights into memory disorders such as **dementia**.* As well as researching HSAM, Robinson leads a team investigating how to detect dementia cases as early as possible, in order to improve outcomes for sufferers.

There are many different types of dementia – but in loss of autobiographical memory, reminiscing may be a beneficial strategy. Robinson believes that greater understanding of why Rebecca has this superior ability to retrieve her personal memories could help to devise specific interventions for people with this type of dementia. While there is still no cure for dementia, there is some evidence that reminiscing about your personal memories helps hold the memories a little bit longer.

Our memories and the stories we tell about our lives are fundamental to who we are. From the conversations we have as children, to the way we narrate events as older adults, the more we nurture our memory, the more precious recollections we'll hold on to.

4

CREATIVITY

Conversations with:

Brittany Harker Martin, Associate Professor, Arts Education and Leadership, University of Calgary, Canada

Professor Nancy Andreasen, Chair of Psychiatry, University of Iowa

David Eagleman, Department of Psychiatry, Stanford University, California

Dr Simon Kyaga, Psychiatrist, researcher; Global Medical Director, Neuropsychiatry, Biogen, Switzerland

Roger Beaty, Assistant Professor of Psychology and Director of the Cognitive Neuroscience of Creativity Lab, Pennsylvania State University

Dr Margaret Webb, Honorary Fellow, School of Psychological Sciences, University of Melbourne

Creativity and good mental health

Creativity is part of us all. Whether we express it through the visual arts, writing, music, science, problem-solving or even our relationships with each other, being creative is what makes us human.

Throughout the ongoing COVID-19 pandemic, there has been a global rise in the number of people experiencing worry, fear and **anxiety**.* Anecdotally, it's been observed that many people, particularly while socially isolated, have engaged in more artistic and creative expression.

Associate Professor Brittany Harker Martin from the University of Calgary has been an arts educator for over 20 years. She's witnessed the mental benefits of creative expression first-hand, and in fact she says that cutting-edge research powerfully supports her observations. The field of art therapy uses art to help sufferers of mental illness therapeutically, but Martin thinks the arts can be used much more widely to promote healthy mental habits.

Mindfulness* is a meditation technique that's increasingly used for managing mental health. It involves paying non-judgmental attention to the present moment. As Martin explains, creativity provides ideal conditions for mindfulness, through a conscious shift into a holistic state of mind often called 'flow', or 'being in the zone'. You are so absorbed with what you're doing that you lose track of time, while feeling fully immersed in a mentally pleasurable, neurochemically rewarding state, which has been proven to reduce levels of the stress hormone cortisol.

Martin acknowledges that we can't be in flow all the time, or we wouldn't function in other ways, but it is a positive mental experience that can be used as a counterbalance to the mental burden our society puts on us. It may not be a silver bullet, but practising creativity can certainly improve our capacity to manage our mental and emotional wellbeing. Now is the time, Martin says, to prioritise creativity in our home lives as well as in the education system.

So, what is it that makes a really creative brain so special? What are the characteristics of extraordinarily creative people that we creatively challenged folk can tap into? I'll tackle that topic next.

The characteristics of creative people

The concept of creativity has captured our imaginations for centuries. What's unique about the extraordinarily gifted minds of artistic or scientific geniuses, and what has allowed them to soar to such creative heights?

The late Neil Simon, author of the award-winning play *The Odd Couple* and a host of other plays and screenplays, had a special genius for finding humour in our everyday lives. He didn't write consciously, more as if a muse were sitting on his shoulder and the words just came flowing out. He reported not knowing what he was going to say until he said it. He described what a psychiatrist would call a kind of dissociative or dreamlike state in which original ideas occurred to him.

The late American theoretical physicist Richard Feynman is the archetypal example of a creative scientist. Somebody could write a very difficult problem on a blackboard and he'd stand there, put his hands on his forehead and think for maybe five minutes, then write out the answer. He would have no idea how he got there; it was just sheer intuition. It might take another two weeks to figure out all the equations that went in between the problem and the solution.

Simon and Feynman were just two of the people interviewed during a study of creativity conducted by world-renowned US neuroscientist and psychiatrist **Nancy Andreasen**. My conversations with her about the creative brain have been truly enlightening.

To get an idea of the character traits of creative people, Andreasen used interviews and personality testing. What emerged from these methods was that creative people tend to be very curious, adventurous and a little iconoclastic. They are sometimes prone to getting into trouble because of their original thinking and rebelliousness. They can be surprised that not everyone sees things that are obvious to them; this is particularly true of creative scientists.

Andreasen also found that highly creative people tend to be obsessive about their work. Michelangelo, she says, is a wonderful example. He often worked for 16 to 20 hours a day, and even rigged up an apparatus consisting of candles that he wore on his head for light, so he could wield his hammer and chisel all night long creating his statues. (Earlier in her career Andreasen was a professor of English Renaissance literature, which explains her fascination with the huge creative breakthroughs during that period of history.)

Andreasen emphasises that creativity is quite separate from intelligence. Some highly intelligent people are creative, but many others are not.

There's also a social dimension to creativity, according to Stanford University neuroscientist **David Eagleman**, co-author of *The Runaway Species: How Creativity Remakes the World*. Creative people are attracted to each other, they work together, and they develop their ideas together. The concept of the creative genius working in isolation is a myth, Eagleman suggests. Artists like Picasso or Van Gogh might have worked by themselves, but just like almost every creator, they were embedded in their social world. Often when people create in isolation, their work doesn't resonate or stick, because they are not getting any feedback from society to tell them what is working for other people.

Nature vs nurture

It seems that creativity partly runs in families, but there are many highly creative people who *don't* have a family history of creativity. Andreasen cites the two great geniuses of the Renaissance, Leonardo da Vinci and Michelangelo, neither of whom came from creative backgrounds. She suggests that there must have been something in their nature, like curiosity, persistence and the ability to see things in a novel way, but they also needed nurture. They were lucky enough to have been born during the Renaissance; Michelangelo also had the good fortune to be practically adopted by the Medici family, and through those connections he was commissioned to do large, high-profile works.

Environmental influences were substantial for both of those artists, and this is important to consider, as the absence of opportunities may result in lost creative excellence. Andreasen feels that one of the most important questions society has to address is how we can help our young children realise their full creative potential. (I'll look further at this issue later in the chapter.)

'Mad geniuses'?

The idea of the mentally troubled artist has become a popular stereotype: think Vincent van Gogh, Edvard Munch, Virginia Woolf, Albert Einstein, Frida Kahlo and Ludwig van Beethoven, just to name a few. But is there a measurable link between genius and insanity?

The archetype of the mad genius is not a concept Nancy Andreasen likes, as she feels it feeds into stigmatisation. She says that while it's true that highly creative people have an increased rate of mental illness, it's more important to ask *why*.

'There are several explanations,' she says. 'One is that if you are one of these creative people, you are always at the cutting edge, and we have a saying in science, if you work at the cutting edge you are going to bleed. And most of these creative people have ideas that aren't obvious to everybody else, particularly when they are young, in high school and college; they get shot down by teachers who don't have the same level of imagination and insight that they do. And that can continue on into their careers.

'And so they've had to suffer criticism, doubt, all kinds of things, and yet the essence of their lives is that they keep going back. They cannot give up because they are so persistent in wanting to pursue whatever their idea is.' A clear challenge for our society is to figure out how to *nurture* this creativity, rather than punishing people for it.

As an English professor, Andreasen observed a high number of creative people with mental illness, particularly **schizophrenia*** (which involves an inability to distinguish between hallucinations and reality). As she moved into the field of psychiatry in the 1970s and 1980s, she led the first extensive scientific study of creative writers. It demonstrated a close association between creativity and mood disorders (such as **depression*** and **bipolar disorder***) – but not schizophrenia, as she'd expected.

Swedish psychiatrist and researcher **Simon Kyaga** led the largest study ever of the association between creativity and psychiatric disorder, involving more than 1 million Swedish patients. The researchers included healthy relatives of patients suffering from these disorders, because they knew from previous studies that even though the relatives don't have the disorder themselves, they still display features of it to a larger extent than the rest of the population. The team's findings were slightly different from Andreasen's: they found a strong link between creativity and bipolar disorder, but also between creativity and schizophrenia.

Many people who are being medicated for a psychiatric condition feel that the treatment hampers their energy and creativity. Kyaga says there have been few studies on this, but there has been *some* research confirming it.

In another study, though, 24 artists with bipolar disorder – involving fluctuations between extreme joy and extreme depression – were asked what their medication, lithium, had meant for their artistic achievements. Six claimed it had made no difference, six said it had been detrimental, and twelve stated it had actually improved their creativity because it had reduced their negative symptoms.

Creativity is not only coming up with ideas, but it's also *doing something* about these ideas. And if you are unable to act on your ideas because of symptoms like depression, you are not being creative. To be effective in your creativity you need to be organised, be able to manoeuvre yourself through the world, and have good social skills to get things happening.

Kyaga adds: 'You do not have to be afraid to be creative, that you might end up getting schizophrenia or some other psychiatric disorder.' As I've already emphasised, creativity can be *really good* for your mental health.

There's another brain condition often linked with creativity that is often viewed as a gift rather than a disorder; it's called **synaesthesia**.* I'll discuss it in detail in Chapter 8.

But what is actually going on in the brain when we're creative?

The neuroscience of creativity

In his research, David Eagleman has examined the essence of human creativity, and how it sets us apart from the rest of the animal species.

The **cerebral cortex*** – the brain's outer layer, where huge amounts of information are processed – is much bigger in humans, which means that there is more room for handling this information. Animals, with their smaller cerebral cortex, essentially act reflexively, whereas humans take in ideas and consider them before deciding whether to respond.

Humans also have a large **prefrontal cortex*** covering the front part of the brain, behind the forehead. This area allows us to think about possible futures and what might happen if we do one thing or another, so we don't have to try lots of things out to see what the consequences are.

But is there an area of the brain specifically for creativity? From her studies of creative people, Nancy Andreasen has concluded that that area is the **association cortex**, a region of the cerebral cortex that provides complex processing of sensory information.

When you read something, your eyes see black and white on the page. The image is sent to your **visual cortex** in the back of your brain, then forwarded to the association cortex, where the black and white letters are turned into words and the words are turned into concepts that have all kinds of associations. Andreasen's theory is that creative people have a more highly developed association cortex, leading to an ability to see associations and connections that others can't. She's shown this using brain scans.

Roger Beaty, Assistant Professor of Psychology at Pennsylvania State University, has also studied the differences between the brains of creative and less creative people. He defines creativity as 'the ability to come up with something new and the ability to come up with something that also fits a given context or goal … So it's a two-part thing.'

He and his colleagues asked subjects to come up with new and unusual uses for an everyday object, answer riddles and brain teasers, and undergo brain scans. Through this testing they identified three distinct brain networks involved in creativity: the **default mode network,*** the **executive control network** and the **salience network**.

The default mode network is activated when you're relaxed and your mind is wandering. Remembering events from the past, thinking about possibilities in the future and daydreaming (which I'll look at in Chapter 6) are all associated with this network.

The executive control network comes into play when you need to focus the content of your thoughts and manage multiple things in your mind at once.

The salience network plays a significant role in detecting and filtering salient information.

Beaty told me that the default mode and executive control networks work separately, so when one is active the other becomes *in*active. The salience network switches between idea *generation*, part of the default mode network, and idea *evaluation*, part of the executive control network. He and his colleagues have found that more-creative people tend to have more network connections, and are more able to exchange information between those networks to produce more original ideas.

Creativity and memory

Beaty is also investigating the complex part that memory plays in creativity.

On one hand, memory is absolutely critical for creativity, because you have to spend years building up knowledge that you can draw upon to come up with new ideas.

However, Beaty says that some evidence shows memory can also *restrict* creativity. If you adhere too closely to what you already know, this can get in the way of coming up with new ideas and moving forward.

There's also some evidence that the way concepts are stored in our memory can influence how we put them together in new ways. Beaty's research is looking at our memory for personal experiences, and how we imagine future experiences.

I asked Beaty whether our creativity changes as we get older. He said that the evidence is mixed. Some research suggests that older people are not as good at creative thinking tasks, some has shown no differences, and some indicates they perform *better* than younger people. Beaty speculates that this may be because they have more information, wisdom and experience to draw on.

How the creative process works

We humans have what David Eagleman calls 'creative software' that makes us restless to keep inventing. This has allowed us to be an incredibly creative species, and to stand on the shoulders of generations before us.

There are many contexts in which we are creative, but Eagleman says what is special about human brains is that they take in information, mix it up in various ways and put out new versions of it all the time. He describes three processes the brain uses to do this: **breaking**, **bending** and **blending**.

Breaking involves taking something that perhaps you're seeing for the first time, and you think about it and imagine what it would be like if it were snapped in half, or had a piece broken off it. It's about breaking things into components to get something else out of them.

An example in the art world is American sculptor Barnett Newman's *Broken Obelisk*. He has created a new way of seeing familiar objects by breaking an obelisk and turning it upside down, so the point balances on top of a pyramid shape. An example in the sciences is DNA sequencing, where long strands of DNA are broken into small parts, making them easier to analyse.

Bending is used to change the size or shape of something. In art, for instance, a human figure might be depicted as a giant, or in miniature. In science, an example is the artificial heart, which functions the same way as an actual heart, except you don't have a pulse; it's a bend in the original idea.

The third creative process the brain uses is blending, which is taking multiple ideas and merging them together. This regularly happens in art. In Picasso's famous painting *Les demoiselles d'Avignon*, he took various styles and combined them in a new way. In the science lab, genetics professor Randy Lewis wanted to make spider silk in large quantities, as it had good commercial potential because of its strength. It's difficult to farm spiders because of their cannibalistic tendencies, so Lewis and his colleagues came up with an innovative idea. They took spider-silk DNA and put it into a female goat, which gave birth to Freckles, a 'spider-goat'. Freckles secretes large quantities of spider silk in her milk.

Eagleman emphasises that it's not just celebrated artists or famous scientists but all of us who use these cognitive processes to create.

Neuroaesthetics

What exactly is going on in our brain when we admire a piece of art?

This is the question that the new field of **neuroaesthetics*** seeks to answer. Biofeedback – monitoring the electrical activity of a person's brain and presenting that information to them on screen – is used to study the effects of visual art on brain circuits.

According to David Eagleman, our brains are attracted to the known and familiar, but at the same time we're hungry for the new and different. Human creativity exists within this exquisite tension.

When you show the brain something new, it has a big burst of activity that can be measured. When it's shown the same thing again and again, that repetition suppresses its activity. This is what causes us to constantly look for the next new thing. But because we also thrive on familiarity, if the next thing is *too* new, it will be disorienting for our brains.

If you produce a piece of visual art or music, or even a new product, that seems exactly like everything that has come before it, it won't get much attention. On the other hand, if you produce something too wacky, which people are not ready for, it won't be received well either. What makes more sense is to tweak the work of others within your culture by adding *some* novelty on top of what is already familiar.

Eagleman uses the example of the Apple iPhone. In 2006, one reporter described it as the 'Jesus phone', because it was so incredibly novel and seemed to have had a virgin birth. However, just like every invention, it actually had a long history: the first touchscreen mobile phone came out in 1993. We tend to ignore the history and forget that nothing comes out of nowhere.

Timing will determine whether ideas fall onto open ears and whether society is prepared to accept them. This is one of the reasons why innovative companies put out a whole range of products, from barely new to wild and extreme. It's a way of figuring out what society is ready for.

How we can nurture our creativity

Our increasing knowledge about the plasticity of the brain supports the notion that we all have the potential to be creative from early childhood right through to our later years. In the early years, there is a lot that our schools could do to help children reach their full creative potential – a topic I'll cover later. As we grow older, we can exercise our brains through exposing ourselves to new experiences and learning opportunities.

Everything that you do in a day is an act of creativity – from working out what you are going to say next or making up a new recipe for dinner, to drawing something or composing a new tune. Eagleman has a few tips to spark our daily creativity.

He suggests it's a good idea to make things if you can, rather than buy them. For example, make a card to give to a friend rather than purchasing one in a shop. If you have children, provide them with toys that encourage creativity like Lego or building blocks, rather than toys where you press a button to make something happen automatically.

Every day Eagleman does something to kick his brain out of its automated routines – for instance, switching his watch to the opposite wrist, or brushing his teeth with his other hand. Most days he finds a new route home from work to avoid getting stuck in a pattern.

He told me: 'You've probably noticed if you drive the same route home from work every day, the first day you do it, it seems to take a long time, and after a while it seems to take less and less time, and this is because you're not really that conscious of it after a while. But if you are doing new things all the time and forcing novelty into your life, things seem to take longer in a useful way.'

Creativity and the 'Aha!' moment

Ever had that experience where you can't seem to solve a problem, then the solution just pops into your head? It's often dubbed the 'Aha!' moment.

While taking a bath, the ancient Greek mathematician Archimedes made the discovery that the volume of water that spilled over the edge, displaced by his body, was equal to his body mass. He ran into the streets shouting *Eureka*!' – 'I've found it!' Other 'Aha!' moments from history include the story of when the apple fell on the young Isaac Newton's head, leading him to develop the theory of gravity, and Francis Crick and James Watson's discovery of the double helix structure of DNA after contemplating the shape of a spiral staircase and comparing it to Rosalind Franklin's famous *Photograph 51*, which revealed the helical nature of DNA.

Dr Margaret Webb is an Honorary Fellow in the University of Melbourne's School of Psychological Sciences. She's passionate about investigating the 'Aha!' moment and its relationship to creativity and problem-solving.

So what is an 'Aha!' moment? Webb says, 'Most people seem to instinctively recognise it – a moment when the lights seem to come on all of a sudden and everything about a problem is clear.

It's usually a very positive feeling and can happen for any problem you might be having.'

Some data indicate that if you are open to new experiences and you go out into your environment and look for new things to try, you are more likely to have 'Aha!' moments. If you tend to see patterns, such as faces in clouds, you are a bit more likely to have 'Aha!' moments as well.

But don't get too excited just yet. Webb suggests that the big creative breakthroughs we usually think of as 'insightful' (like Gutenberg's printing press and Archimedes' Eureka moment in the bath) are slow, gradual processes of hard work. Moments of insight are definite drivers towards creativity, but you must also work in a focused and incremental way towards your goal. You still need the skills and capabilities to make a revolutionary discovery or create a masterpiece.

Webb says 'Aha!' moments are usually associated with quite positive emotions. But not always. One of Webb's friends has coined the term 'D'oh' moment to describe a sudden insight where you realise you've been doing the wrong thing (like the moment you realise you're at the wrong airport). You can feel a bit stupid, but this might bring a new pathway for the solution and energy for solving it.

The Aha! Challenge

Margaret Webb and her colleague Dr Simon Cropper have been part of an international research collaboration aiming to measure people's creative thinking and potential.

In 2019, in collaboration with the ABC, they launched The Aha! Challenge, a large-scale citizen-science project that was the first of its kind. Around 15,000 people of all ages, from all walks of life and from more than 100 countries, took part.

Webb described to me some of the tasks they were asked to do, which included word puzzles: 'You're given three words. So, for example, if I gave you the terms potato, heart and tooth, and then the task is to find the word that meaningfully combines with each of those words. So in this case it would be "sweet". So you've got sweet potato and sweetheart and sweet tooth.' It's a commonly used problem type, but because it's so verbal, that can skew the way people can answer. Another task is the matchstick arithmetic puzzle, in which people are asked to solve an equation written out in Roman numerals using matchsticks, just by moving one matchstick.

Another task Webb and her colleagues created for The Aha! Challenge was the 'Divergent Association Task', whereby participants have to list 10 of the most unrelated nouns they can think of. Margaret adds, 'Since we published our findings on it, it has gone viral on twitter and YouTube. We currently have responses from almost every country in the world.'

Webb admits that measuring these 'Aha!' moments is still quite subjective, but there are some physiological and behavioural reactions that can be tracked. For example, certain changes in emotional states can be detected through people's galvanic skin response, a change in the electrical resistance of the skin measured using a galvanometer. One interesting emotional response which has been detected is that people tend to laugh when they have an 'Aha!' moment. There's very little data on this at the moment, making it a new and exciting field.

Research into creativity, problem-solving and the 'Aha!' moment could be valuable to society in several ways. Webb points to the way the 'Aha!' moment is associated with motivation. Education Professor Peter Liljedahl, based at Simon Fraser University in Canada, reported in a 2005 paper on his investigation into a group of people who didn't want to learn mathematics but had to do it as part of their course, and whether they had 'Aha!' moments or not. Those who did have 'Aha!' moments became enthusiastic about mathematics after a while because they realised that it could be fun, and they could be competent at it. He concluded that having these 'Aha!' moments contributed to their motivation. Webb says studies such as those run by Amory Danek and her colleagues (2013) and Jasmin Kizilirmak and her colleagues (2015) have also shown that 'Aha!' moments are associated with improved memory and learning.

'There have been some papers investigating increasing the "Aha!" moments that people have,' says Webb, 'and there are things like looking for every possible answer for a situation, and following those through, not just the typical answer you think it is. And also looking for contrary instructions. So if this was the opposite, what would that be? And then these sorts of things can increase "Aha!" moments.'

Webb told me about some human findings of The Aha! Challenge project. 'Aha!' moments range from life-changing epiphanies about the nature of existence to everyday realisations. They're common when weighing up questions and issues in our lives, relationships and work – such as how to balance conflicting calendar appointments or points of view. A number of individuals who completed the survey noted that their 'Aha!' experiences came when thinking about how their children, or their parents, were viewing a problem.

'Aha!' experiences are common in learning – many people reported 'Aha!' moments during study of physics, maths or a foreign language, particularly when they realised that seemingly disparate fields or solutions were connected. You can experience 'Aha!' moments when you do the weekly crossword or word puzzle, or even when you solve a mystery in a TV series or book before the solution is revealed. But 'Aha!' moments can also be fairly mundane realisations – like figuring out how to fix the vacuum cleaner or the car, or how to set up a smartphone, or how to reorganise the house, how to unblock a sink, and so on.

Moments of insight often occur when people are under stress, for example when having to solve critical and urgent problems very fast and usually alone. In terms of where they occur, 'Aha!' moments are most frequently reported in nature, in or near bed (waking or sleeping), or otherwise in a quiet place where the brain can just tick over: while driving, doing housework, in the shower, on a walk, knitting, or waiting for sleep.

Teaching creativity

As I've mentioned, everyone has the capacity to be creative; it's part of the human condition. However, not all of us will realise our full potential, and that greatly depends on the quality of our education. There's a lot more that could be done to nurture creativity in the young, both outside the school system and within the curriculum.

Nancy Andreasen feels that schools often set young people on a particular narrow track and it becomes difficult to venture off it. She suggests it's a mistake to require young people to decide too early whether they are going to pursue a scientific or artistic career path. She says the more they can learn about multiple topics, the richer their brains will be, and the greater their opportunity to see things in original and creative ways.

Roger Beaty would also be interested in further research on whether our creativity can be improved with training. There's a lot more to be known, he thinks, about how we can teach for creativity in the arts and sciences, and to what extent this kind of teaching can change the wiring of the brain for creativity.

David Eagleman says that unfortunately children's opportunities to realise their creative potential usually relate to their socio-economic status. If you are in a high socio-economic class, have parents who cultivate and encourage your creativity, and are given good-quality schooling, Eagleman says you have a much higher chance of succeeding creatively. He feels this is a tragedy, a waste of what is so special about the human brain.

Unfortunately, in impoverished neighbourhoods, schools don't often prioritise the arts. What is so special about the arts is the way they teach children how to use creative processes, rather than just focusing on getting the answers right in tests.

Eagleman adds that turning schools into more creative places is easy, and doesn't have to affect the school budget: 'A teacher's job really is to teach what is known, and then by the time you get near the end of the semester, to use that as a springboard for students to mash things up and make their own versions of things.'

In an article co-written with colleague Mitchell Colp, Brittany Harker Martin presents art as a way to promote mental health in schools. Schools need to schedule time for visual and tactile activities that will interrupt the cognitive intensity of the rest of the day.

Martin and Colp asked participants to do simple, art-based exercises for 20 minutes daily for a week. They tested them before and after the exercises, and found that by the end of the week their mental health scores had increased.

Martin suggests that people are easily drawn to the visual and tactile processes of smartphone use because our brains are craving that type of interaction. There are healthy mental habits to adopt, including visual art activity, and there are unhealthy mental habits, such as spending hours on social media platforms and mindlessly consuming content often posted with commercialised intent.

Creativity is the future

Eagleman is optimistic about the ability of the human species to create a bright and stimulating future. In his opinion, the main thing that needs to change to make it even brighter is to improve the schooling system, so that children's natural creativity is encouraged and nurtured.

Creativity is the cornerstone of our next steps into the future, according to Eagleman. Because we are getting better and better at artificial intelligence, many of today's jobs will not exist in 10 or 20 years' time. The world is changing so rapidly that the skills children are being taught may not be as relevant as being able to process new data in a creative way.

Rather than train our children in the way we have done since the Industrial Revolution, by sitting them down and having them memorise and learn by rote, the thing we need to emphasise in the classroom is creativity.

5

SLEEP

Conversations with:

Guy Leschziner, Professor of Neurology and Sleep Medicine,
King's College, London

Professor Penny Lewis, School of Psychology, Cardiff University

Sylvain Williams, Professor of Psychiatry,
McGill University, Montreal

Professor Danny Eckert, Director, Adelaide Institute
for Sleep Health, Flinders University

The mysterious world of sleep

Jackie always knew she was a sleepwalker, even during childhood. But as a young adult, she was shocked one morning when her landlady asked her: 'Where were you going on your motorbike at one in the morning?'

As far as Jackie knew, she'd been soundly asleep in bed all night. But it soon became clear she'd actually been riding her motorbike in her sleep! Her response was to sell the motorbike very quickly, in case she did the same thing again.

Decades later, when she was in her 70s, neighbours witnessed her driving along the seafront in her car in the middle of the night. Again, she was completely unaware of this hazardous behaviour.

* * *

Dr Guy Leschziner is Jackie's neurologist (a specialist in brain and nervous system disorders). He's fascinated by the field of sleep medicine, because he believes it gives us insights into the various states of our brain. In his book *The Nocturnal Brain* he explores the neuroscience behind our sleep, and he told me about some of the bizarre ways in which our sleep can become disordered.

Sleepwalking is one of the best known, and can lead people to put themselves in real peril – jumping out of a second-storey window, sexual behaviour that gets them into terrible trouble, or, as we saw in Jackie's case, riding a bike or driving a car in the middle of the night.

Other disorders include sleep talking and sleep eating. Leschziner has seen case reports of people who've drunk bleach or eaten raw meat and other foods that they would never consider consuming during the day.

For people like these, falling asleep must feel like diving into danger – but for me, sleep has always been a welcoming place, known as Blanket Bay.

That's the name my mother always used to lure me to bed as a child. There's a real 'Blanket Bay' on the Great Ocean Road in Victoria, and another one in New Zealand, where European pioneers sheared their sheep under shelters stitched together from blankets. To Mum and me, though, Blanket Bay signified the silent and mysterious world of sleep.

We humans spend a third of our lives in this state – more time than we spend on most other things – so it must be important to us. We all know what happens when we don't get enough of it. I know I get cranky, my mood dips, I crave junk food, I'm highly emotional and I just want to escape back into Blanket Bay.

But what is sleep for, what is happening in our brain when we're asleep, and how does it lead to the strange disorders I've just looked at? Come with me to 'Blanket Bay' now and find out.

Why do we sleep?

I put this question to **Professor Penny Lewis**, who works at the Neuroscience and Psychology of Sleep Lab at Cardiff University. 'This is really the million-dollar question,' she replied.

Scientists have been trying to work out the functions of sleep for more than a century, and in the last 20 years an explosion of research has shown that sleep is crucial to many of the processes happening in our brains and bodies.

'We do know that it's important for things like your immune system,' says Lewis: 'It's important for helping your body to maintain the correct temperature, and there are all kinds of negative side-effects to sleep deprivation.'

Studies have shown that if rats are deprived of sleep for long periods of time, they die. In humans there is a rare hereditary brain disorder called fatal familial insomnia that can also eventually lead to death. Disturbingly, sleep deprivation has been used as a means of interrogation and torture. So sleep is clearly crucial in order for us to survive and thrive.

I asked Lewis: what's actually happening in the brain when we sleep?

'So, anecdotally many people seem to think that the brain just switches off when we are asleep because maybe that's what it feels like to us,' she said. But 'instead it moves through an incredibly highly structured series of very specific sleep stages'. She explained that 'each of these stages is characterised by different kinds of brain activity, and also different concentrations of chemicals in the brain. We think that each of these stages is important for different things and that they act together to build up or to support the overall functions of sleep.'

There are two types of sleep: **rapid eye movement (REM) sleep** and **non-REM sleep**. Non-REM sleep has **three stages**. After moving through quiet wakefulness in stages one and two, you reach stage three, the deepest stage of sleep, when the waves of electrical activity in your brain are at their slowest. If you were to be woken during this stage, you would feel disoriented for quite a few minutes. Your brain takes about 90 minutes to cycle through non-REM stages one, two and three, and then REM, and it does this several times throughout the night. Non-REM sleep periods are usually longer in the beginning of the night, whereas REM periods are longer later in the night.

The REM stage has some unusual characteristics, which I'll look at in more detail later.

Sleep and memory

Lewis is particularly interested in the relationship between sleep stages and the way new information and recent experiences are turned into long-term memories.

Neuroscientists think this involves the process of **rehearsal**,* where things you do during the day that are important to you or hold some emotion for you are spontaneously replayed the next time you sleep. Like rehearsing before a performance, this strengthens the connections between the neurons involved in what happened before you went to sleep.

One compelling study involved getting people to play a simple card game that you may be familiar with. You lay out an array of cards face down, and every card has a match. You have to flip two cards over, and if they match you take those two cards away. The idea is to try to remember where particular cards are so that you can pick matches as the game progresses.

In this study, as the players were learning the locations of the cards, they were given the scent of roses to smell. After a while they were tested to see how well they could pick up matched cards. Then they were told to go away and get a good night's sleep. They came back the next day and were tested on the same card layout.

Different groups were then given the smell of roses again, but at different times. One group were given the smell when they were awake, one during REM sleep and one during deep non-REM sleep. It turned out that the group who smelt the roses again during deep sleep performed much better the next day than any of the other groups. The researchers concluded that this was because deep sleep is important for strengthening memories.

Lewis points to another study that suggests it's possible to increase the amount of deep, slow-wave sleep we get, and in turn strengthen memories. This can be done by passing an electrical current across the **cerebral cortex*** at the frequency of those slow waves, which triggers more slow waves than you would otherwise have.

A more user-friendly technique using sounds is being explored too. If tones are played softly at the frequency of slow waves, our brainwaves will synchronise themselves to this frequency, and this has been shown to improve memory consolidation.

Lewis and her colleagues are excited about this research, as it may mean you could improve your memory and learning by controlling your sleep. These new techniques may also be able to help people with sleep disorders like the ones I'll discuss later in the chapter.

REM sleep, the 'dark horse'

Lewis refers to the REM stage of sleep as a bit of a 'dark horse' when it comes to sleep research, as there are still so many disagreements and unanswered questions – yet its discovery in 1953 by US dream researchers Nathaniel Kleitman and Eugene Aserinsky is seen by many scientists as marking the beginning of modern sleep research.

The REM stage of sleep occurs at the end of the 90-minute sleep cycle, following deep sleep. Suddenly our brain starts to become more active, similar to the pattern it displays when we're awake. Our heart rate increases, our breathing becomes faster, and our eyes flutter rapidly under closed lids (hence the name 'rapid eye movement sleep').

REM sleep is often known as dream sleep, though it's now believed that we dream in the other sleep stages too. During REM the body becomes immobilised, which is known as muscle atonia. Sleep researchers think this is to prevent us from acting out what happens in our dreams. (I'll look at dreaming in more detail in the next chapter.)

I've mentioned that deep sleep is important for memory and learning – but REM sleep is important for those things too.

Of mice and men

Sylvain Williams, Professor of Psychiatry at McGill University in Canada, runs a lab investigating memory consolidation during sleep. He explained to me that REM sleep is a difficult phase of sleep to study in humans, but in an article published in 2016 he and his colleagues showed for the first time that REM sleep is involved in memory formation in mice.

Studies had already shown that if you disrupt REM sleep, the animal doesn't perform adequately the next day. However, such studies are difficult to do, because you have to wake the mice up precisely during REM sleep, which varies in length, and the process is also very stressful for the animals.

So, Williams and his team used a powerful neuroscientific technique called **optogenetics**. It involves the use of light to control the activity of neurons that have been genetically modified to become more light-sensitive. The neurons can then be either inhibited (slowed down) or excited (sped up), simply using light.

With this technology they were able to specifically disturb REM sleep without disrupting non-REM sleep. After waking the mice up, they tested them on their memory of the location of objects in their environment. They also tested a second group of mice, the control group, whose REM sleep had *not* been disturbed. They found strong evidence that the memories of the animals whose REM sleep had been disturbed was significantly worse.

Williams acknowledges that the same type of experiment cannot be done on humans, but he notes that REM sleep in humans has very strong similarities to REM sleep in rodents – so this is good evidence that the same could be true in humans.

What's more, there is other evidence that REM sleep is important for certain types of memory in humans. There have been studies in which people went to sleep after learning lists of word pairs. During sleep their brainwaves were monitored, and every time they went into REM sleep, the researchers woke them up. The study found that when they were woken up there were disturbances in their memory of the word pairs.

By the way, I don't know about you, but this got me wondering whether mice dream during REM sleep! Williams says it's hard to say – although there *is* evidence that mice engage in **rehearsal**,* the process I looked at earlier in the chapter: when they sleep, they practise how to get to where their food is. The team are interested in investigating this in their lab in future.

'Sleep on it'

In the previous chapter on creativity I discussed the 'Aha!' moment. Here's a variation. How often do you wake up and suddenly realise you've come up with a solution to a problem you were grappling with, or made a clear decision about something you were unsure about the day before?

Remember the old adage 'go sleep on it'? Penny Lewis says it's fantastic advice. There's been substantial evidence in recent years to suggest that sleep can assist us with both problem-solving and decision-making. Sleep helps us not only strengthen memories but also retrieve specific information from a set of memories, providing an overview – which may lead to a better solution or more sensible decision.

'So an example of a case where sleep has been shown to help with tricky problem-solving comes from a task that we call the rapid association task,' Lewis told me. It's the same task used in the Aha! Challenge, which I looked at in the last chapter: asking people to find the association between three different words.

Two groups of people are given a number of these problems and their performance is measured. Then they are allowed to have a break of a couple of hours, with one group sleeping during that time and the other group not sleeping. The group who sleep, particularly if they've had REM sleep in that period, seem better able to find the answer.

Using data from this type of creative problem-solving experiment, Lewis and her team have developed a model to explain how REM and non-REM sleep might assist problem-solving and decision-making in different but complementary ways. The model suggests that non-REM sleep helps to organise information into categories, and REM sleep allows us to see unexpected connections beyond those categories. It's thought that replaying this information in our minds during sleep allows us to make these linkages.

REM sleep and emotions

REM sleep is also believed to have an important role in processing emotional memories. 'So there's an idea that if emotional memories are replayed during REM sleep,' says Lewis, 'then that may help to sort of extract some of the emotional content, so that when you think of those things next time when you are awake they are less upsetting.'

Lewis told me that while there's a bit of controversy in this area, one popular and compelling idea – developed by Matthew Walker, Professor of Neuroscience and Psychology at the University of California, Berkeley – has been labelled the **overnight therapy** hypothesis.

During REM sleep, our physiological state is different from when we're awake. The stress chemical noradrenaline is at a very low level, which means that our heart rate and other measures of emotional arousal can't increase – creating a stress-free environment. Walker's idea is that if you replay upsetting or traumatic emotional memories during REM sleep, this will strengthen the content of the memory, but it may also weaken the emotionality of that memory, because you won't be experiencing those physiological responses as you're replaying it.

The result could be that what you remember when you wake up is not as upsetting as it was prior to falling asleep. This may be an important part of how our emotional system resets itself and deals with unpleasant things.

When REM sleep is disrupted

Obstructive sleep apnoea (OSA)* is a condition that can reduce or disrupt REM sleep, according to **Professor Danny Eckert**, now Director of the Adelaide Institute for Sleep Health at Flinders University.

OSA is very common, affecting approximately 1 billion adults worldwide. It's a condition in which the throat narrows, or closes over completely, up to 100 times per hour during sleep. This means you can't get enough air into your lungs, and breathing becomes very difficult. There are other increased health risks, such as hypertension and cardiovascular problems, as well as impairment of glucose function in people with diabetes. It's been found that OSA is most severe when it occurs during REM sleep.

The main treatment for OSA is continuous positive airway pressure (CPAP) therapy, which was developed through the pioneering work of Colin Sullivan, Professor of Medicine at the University of Sydney. The patient wears a mask to bed, and just enough air is blown through the mask to prevent airway closure during sleep. 'However, the problem is half the people that try CPAP therapy are unable to tolerate it,' says Eckert.

A recent study in New York tested OSA sufferers by giving them CPAP therapy one night, then taking it away another night during REM sleep only. 'And before they went to sleep in each case they gave them a learning task,' Eckert explains. The researchers found that 'when the REM sleep apnoea was removed with CPAP therapy, the learning was actually enhanced'.

'So now that we've understood more about exactly why people get sleep apnoea, it's a really exciting time for the field.'

Eckert says these new discoveries about sleep apnoea in REM sleep, and some of the other stages of sleep where muscle activity changes, are offering important insights and potential pathways for developing new and effective therapies, and a better understanding of why people get sleep apnoea.

Strange disorders of sleep

Remember Guy Leschziner's patient Jackie, and her dangerous sleepwalking? **Parasomnia*** is the term for this and other unusual behaviours that some people experience during REM sleep, non-REM sleep, or the period between sleep and wakefulness.

Leschziner told me scientists now have evidence that behaviours like sleepwalking occur when different parts of the brain are in different stages of sleep or wakefulness. During these experiences, the parts of the brain responsible for rational thinking, planning or aspects of our personality often remain in very deep sleep. (This can be observed through the electrical activity of the **frontal lobes**, which are responsible for these functions, as well as the **hippocampus,*** where the majority of our memory resides.) Meanwhile, the parts of the brain responsible for movement and vision show the activity of wakefulness.

It's this disjointedness of the brain that results in parasomnia.

One to two per cent of adults have these sleep experiences, and it's thought that some of these people have a genetic predisposition. But some conditions that disrupt the deep, slow-wave sleep stage – like **OSA,*** snoring and **periodic limb movement disorder** (a condition that causes involuntary kicking at night) can also make these sleep experiences more likely.

Insomnia

Most of us have experienced the frustration of not being able to get to sleep, or waking up at some point in the night. It's the most common sleep disorder; about one-third of adults have intermittent insomnia symptoms and about 10 per cent have chronic insomnia.

This disorder can significantly affect your life, leaving you tired and lethargic during the day, which can have an impact on your overall health, relationships and productivity.

Risk factors include stress, **anxiety**,* **depression**,* obesity and a range of medical conditions, as well as stimulants such as tea, coffee and alcohol and some over-the-counter medications. Females are twice as likely to experience insomnia as males.

It can be caused by other sleep disorders, but can also *cause* other sleep disorders, because it increases our drive to remain in very deep sleep. Normally somebody will wake up fully if their deep sleep is disrupted, but if you are sleep-deprived that doesn't tend to occur.

Sleep hallucinations

Some people have unnerving sleep-related experiences that actually occur while they are awake. They're called **hypnagogic hallucinations**.

Typically, a person is drifting off to sleep, then wakes fully but can see things in the room, like a dark figure standing over the edge of their bed, or bizarre animals in their room.

(I'll look at other types of hallucinations in Chapters 10 and 16.)

Sleep paralysis

Related to these hallucinations is **sleep paralysis**, one of the more common parasomnias. The sufferer feels fully awake yet is completely paralysed. Between 20 and 40 per cent of the population report they've had sleep paralysis at least once in their life. A smaller percentage of people experience related symptoms, like difficulty breathing, or (similar to hypnagogic hallucinations) sensing a presence nearby.

'So you can imagine how terrifying that is,' says Leschziner, 'to feel completely paralysed and to feel that there is a stranger in the room standing over you.' He explains that 'these phenomena arise when there is a blurring of the boundaries between rapid eye movement or dreaming sleep and wakefulness. The hallucinations that we experience really represent some of that dreaming process kicking in while we are fully awake. And if those aspects of REM-like paralysis also kick in at the same time, that results in sleep paralysis.'

Sleep paralysis doesn't often need treatment, but if you do have this experience and it's preventing you from getting a good night's sleep, it's recommended you consult a doctor.

Narcolepsy and cataplexy

Narcolepsy is a rare sleep disorder that causes a person to suddenly fall asleep during the day, or at inappropriate times. Leschziner has known people to fall asleep playing football or even while having sex. Again, it can also cause hallucinations and sleep paralysis, because people go from wakefulness directly into REM sleep, which doesn't normally happen.

Leschziner adds that as part of narcolepsy a condition called 'cataplexy' is also seen. This is where sudden strong emotions such as laughter, fear, anger, stress or excitement while a person is awake can trigger loss of voluntary muscle control, like the paralysis we normally associate with REM sleep. He gives an example of a person who might be about to deliver the punchline of a joke, and, when the REM sleep paralysis mechanism is triggered, they collapse to the ground.

Narcolepsy results from the damage to a very small number of neurons deep in the brain in an area called the **lateral hypothalamus,** which is fundamentally important to sleep regulation, and to our ability to stay awake and switch between REM and non-REM sleep. 'And we think that this damage is really as a result of autoimmune attack, that people who have an underlying genetic predisposition who are exposed to a particular virus or other pathogen in the environment trigger their immune system to attack this very small part of the brain,' Leschziner explains.

When Leschziner was at medical school, narcolepsy was poorly understood and often the subject of jokes. However, his fascination with the condition has led him to run a large clinic for people with the disorder – because every day he sees the terrible devastation narcolepsy causes people in most aspects of their lives.

Although treatments are being investigated, for the moment there is no cure. Scientists are trying strategies focusing on suppressing the immune system, stem cell transplants and gene replacement therapy. This research is still in its infancy.

Acting out dreams

One of Leschziner's patients, John, has a REM sleep behaviour disorder that means he acts out his dreams. He has had very vivid, frightening dreams of being chased or attacked by large animals. He'll often kick out or grab his wife or swear in the middle of the night.

As I've mentioned, usually in REM sleep we're paralysed; the only muscles that move are the eyes and the muscles involved in breathing. However, in John and others with this disorder, the mechanisms that normally generate paralysis in REM stop working, allowing them to act out their dreams.

Treating parasomnias

So, what is the value of studying people with these extreme sleep disorders?

'We understand about the processes of sleep and what can go wrong,' Leschziner told me. 'And so we get some insights into how the normal process of sleep is regulated by the different mechanisms of the brain. Essentially they give us a window into normal sleep.' Such insights can only benefit us all.

Leschziner has found that the most effective treatment for many of these problems is to deal with the daytime stress or anxiety that may be causing them. There are also some drug treatments that reduce the likelihood of these behaviours. Drugs are certainly recommended by Leschziner for some of the extreme cases that are referred to his clinic, where people put themselves in real danger.

Leschziner emphasises that good sleep is essential for us to survive and thrive. We now know that if you sleep poorly, you are putting yourself at greater risk of cardiovascular disease, high blood pressure, stroke and **Alzheimer's disease.*** Lack of sleep can also contribute to a decline in mental health and wellbeing.

However, Leschziner believes there's a balance to be maintained.

He frequently sees individuals who have read or heard about the risks of sleep deprivation and become preoccupied by the fact that they are not sleeping well. When these people are studied in clinics, their sleep is not all that bad, but the stress and anxiety they've developed about sleeplessness can then generate severe insomnia.

You don't necessarily have to get the recommended seven to eight hours' sleep a night. There are huge differences in individual sleep needs, often determined by genetics and other factors. If your sleeping and waking times are regular, and you generally wake up feeling refreshed, you are probably getting enough good-quality sleep. It's best to be guided by how you feel, Leschziner says.

So, what are you waiting for? Best wishes for a great night's sleep – enjoy Blanket Bay and sweet dreams!

6

DREAMS

Conversations with:

James Pagel, Associate Clinical Professor of Family Medicine, University of Colorado

Paul Davies, Regents Professor, Department of Physics and Beyond Center, Arizona State University

Dr Stephen LaBerge, psychophysiologist

Dr Susan Long, Research Lead and PhD Program Co-Lead, National Institute of Organisation Dynamics Australia

Franca Fubini, psychoanalytic psychotherapist, organisational consultant, group analyst and Chair of the Social Dreaming International Network

Muireann Irish, Professor of Cognitive Neuroscience, School of Psychology and Brain and Mind Centre, University of Sydney

Why do we dream?

For as long as I can remember, I've experienced a vivid dream life. Often, at the breakfast table, I've begun the colourful saga of the dreamworld I've just emerged from: endlessly fascinating – according to me!

Usually I'll quickly sense the eye-rolling and bored yawning coming from my family: 'Here she goes again!' Mmm, perhaps not so interesting to others – after all, our dreams are unique and intensely personal.

But why do we dream, what do our dreams mean, and do they matter? These are questions that have been grappled with for millennia, by scientists, philosophers, psychologists, poets and artists … so let's explore.

In ancient times, dreams were a central part of religious and metaphysical belief systems and considered to have supernatural and predictive powers. The earliest known recorded dreams were written onto clay tablets in Mesopotamia around 5000 years ago.

By the 18th and 19th centuries, philosophers had begun to consider dreams more rationally, moving away from supernatural explanations to an understanding that they come from within the dreamer's mind.

From the mid-1850s, a more medical and scientific approach to the study of dreams emerged, with researchers wanting to understand what happens in the brain when we sleep, and what makes dreaming distinct from thinking.

In 1900, the founder of psychoanalysis, Sigmund Freud, released his seminal book on dream theory, *The Interpretation of Dreams*. Suddenly the concept of dream analysis became extremely popular. Freud believed that dreams are fulfilments of repressed wishes, and that dream content includes symbols representing the dreamer's unfulfilled desires. He developed dream interpretation within psychoanalysis as a way of tapping into unconscious thoughts and feelings.

Since Freud, there's been a wide range of theories on what the function of dreaming is. They include:

- **Memory consolidation** – dreams are replays of past events and we consolidate those memories while we sleep. (I looked at this idea in the last chapter.)
- **Emotional regulation** – dreams emerge from our emotional lives while we are awake, and they could help us regulate and process our emotions while sleeping. (Again, this was mentioned in the previous chapter.)

- **Threat simulation theory** – dreams provide a virtual world in which we can practise survival skills and overcome threats.
- **Social simulation theory** – dreaming allows us to rehearse social interactions and perceptions so we perform better in our waking lives. (I looked at the idea of **rehearsal*** in the last chapter.)
- **Activation synthesis** – dreams are random strings of memories that may prompt us to make new connections and come up with creative ideas while asleep. (I'll look at the link between dreams and creativity below.)
- **Empathy theory** – dreams only have a function when we share them with others. Sharing stories and dreams can help to build empathy between people. (Later I'll look at **social dreaming,*** which is a similar idea.)

These are just some of the recent lines of thinking on dreams, but this is a field of study that is rapidly evolving ...

The new science of dreaming

Our understanding of dreams is currently undergoing a dramatic shift.

According to **Associate Professor James Pagel** of the University of Colorado, who has studied sleep and dreams for over 40 years, the advent of brain imaging is fundamentally changing our conception of sleep.

It's thought that we dream for at least two-thirds of the night, if not more. However, we're likely to remember perhaps only 5 per cent of the dreams we have: the average adult only recalls about four to six dreams a month, and even those seem to dissipate very quickly unless we write them down (more on that later). So, whatever function dreams serve, it doesn't seem to rely on our remembering them!

More recently it's been discovered that the *type* of dream you have varies according to the stage of sleep you are in.

In the past it was widely believed that dreams only occurred during the REM phase, but Pagel says REM sleep can occur without dreaming, and dreaming can occur without REM sleep; many researchers now believe that we dream during *all* stages of sleep.

The dreams we have **when we're falling asleep** are called **hypnagogic dreams**. They are very visual and intense – almost like the **hypnagogic hallucinations** I looked at in the last chapter – but they lack a narrative.

The dreams we have in **light sleep** are often rambling, unfocused regurgitations of the day's waking activities.

The dreams we have during **REM sleep** are the most vivid and bizarre, and the ones we are most likely to remember. They tend to be long, story-like narratives that often closely resemble the waking state (just as REM brainwave patterns are similar to those of wakefulness). The dreams of **deep sleep** come from deep in the mind. Says Pagel, 'The dreams of deep sleep are what some people call dreamless sleep, or these incredibly bizarre, strange events that sometimes are more like sleepwalking: confusional arousals, night terrors, extreme emotions, extreme body sensations and intense but very undeveloped thought processes.'

Pagel describes dreams as 'tunnelling between the body and the mind'. They consist of visual components, memories and emotions – all of which have *biological* markers. But there are other features of dreams that are *mind*-based. Pagel says we use dreams in the creative process, and they give us a way of understanding that's different from conscious thought.

Pagel is greatly intrigued by this link between dreaming and creativity. He had conducted studies on successful creative individuals, but noticed that not much research had been done on larger groups. He decided to investigate how people use their dreams, by distributing a questionnaire among a general population in Hawaii. He and his team found that most people in this population used their dreams in their relationships, in decision-making, and in responding to stress in their waking behaviour.

To further explore the connection between dreams and creativity, Pagel conducted a survey of film directors, screenwriters and actors working at the Sundance Institute in Utah. He was amazed at how they used their dreams in their work, and noted a difference based on each individual's creative interest. Directors used their dreams in responding to change in stress, screenwriters used them in decision-making, and actors just used their dreams across the board. He says that no one uses their dreams like an actor!

Pagel and his team then used the same survey in their sleep lab, rating their patients' creative interests. They found that if individuals have a creative interest, no matter what it is, they actually use their dreams to enhance it.

Getting reports from people about their dreams, along with the use of the latest brain imaging technologies, is gradually giving us more insight into the dreaming brain. When I asked Pagel about the future of dream research, he was excited to say that the possibilities are wide open.

One of the things he's been looking at is whether artificial intelligence systems are already dreaming. He even goes as far as to say that there's better evidence that machines are dreaming than there is that non-human animals are dreaming! Dreaming machines – it's a space to keep an eye on!

Remembering your dreams

How can we learn to recall our dreams?

It's often said that practising meditation or **mindfulness*** is a good approach, as improving our ability to focus our attention inwardly can help us to better understand our thoughts and emotions and, in turn, increase awareness of dreams. Another tip is to try keeping a **dream journal** by your bed – and when you're ready to go to sleep, tell yourself several times that you will remember your dreams in the morning. Give yourself a chance before you get up to think about your dreams and write things down, even if they are just vague ideas or snippets of images.

Then ask yourself questions about your dreams. How did you feel during the dream? What did the setting and the people remind you of?

You can then explore whether the dream, or particular aspects of it, remind you of ongoing concerns in your life – and what that suggests about who you are and how you interact with the world.

I've kept a dream journal by my bed at various times in my life – and I have to write them down very quickly, to capture the detail before it's gone forever. I find myself marvelling at the detail and intricacy of the images, events and characters appearing in what feels like the theatre of my mind.

It's not so marvellous, though, when the dreams are nightmares, filling me with fear and horror. When that happens, there's nothing like the sense of relief as I slowly emerge into the reality of wakefulness to find that it was all just a bad dream.

But there's another dream state some of us are lucky enough to experience, in which remembering your dreams can be *even more* important.

Lucid dreaming

Have you ever had a dream in which you're *aware* that you're dreaming – like a dream within a dream?

This seemingly altered state of consciousness is called **lucid dreaming.*** Such dreams can be disturbing if you're not sure what they are, but many lucid dreamers actually have the ability to take control of their own dream narratives.

'The thing about lucid dreams,' says theoretical physicist **Paul Davies**, 'is that it's not like the real world where you are constrained by all sorts of things, including the laws of physics; you can do magic.

'I began lucid-dreaming when I was a teenager. I'd always suffered from nightmares, even as a very young child, and this was very terrifying. But in my mid to late teens I had what at the time was a very disturbing if not terrifying different type of dream in which I thought I was awake but actually I was asleep.

'It seemed so vivid, everything seemed so real and normal, I could see the bedroom around me, and yet I felt paralysed. I felt I was trapped in sleep and that this might very well go on forever. And then I realised in later years that these were what people called lucid dreams, and then once I realised that this was a particular type of dream state I began to get interested in it and began to try to induce these lucid dreams.'

He explains that 'a lucid dream is not just the same as a vivid dream – everybody has vivid dreams. A lucid dream is where you can have all of the fine detail – so for example you look at a tree and you can see individual leaves in all their glory – and you have the tactile sensation. You get the whole range of sensory experience.'

Once you discover you're in this dream state, he says, that's when the fun starts. He admits that it can have a scary side if you don't know what's happening, but an adventurous side once you gain control over it.

I can't conclusively say whether I've had a lucid dream experience or not. Lucidity is not always a clearly defined state. I've had some very real and vivid dreams that felt different from my regular dreams, but I'm obviously not a proficient lucid dreamer.

A little over half the population report having had a lucid dream – and nearly a quarter of us have the experience once a month or more. Less than 1 per cent of the population are highly skilled and regular lucid dreamers, and almost all of them have exceptional dream recall. In fact, some say the strongest predictor of whether you have lucid dreams is how well you remember your dreams generally.

How do you do it?

Dr Stephen LaBerge is a psychophysiologist (specialist in the interaction between the mind and body), and a world-renowned expert on lucid dreaming. He pioneered the scientific study of this dream state at Stanford University in California, and has spent decades researching it. He explains that lucid dreaming occurs during REM sleep and usually lasts for two or three minutes, but sometimes for up to an hour.

LaBerge's interest in lucid dreaming began when he had his first experiences of it as a student in the late 1970s. So, in order to study this state, he taught himself to dream lucidly at will. Now he's found ways to teach others.

So, how is it done? 'Well,' he says, 'you can learn to have lucid dreams basically by developing your dream awareness, meaning starting with better and better dream recall, so you remember most of the dreams of the night.' This is where dream journalling and the other dream recall techniques I looked at earlier come in.

'And then you come to recognise what kinds of elements of your dream could tell you that you are dreaming, things that are odd that are characteristic of dreams, that when they happen, if you noticed you could become lucid. And these we call dream signs ... A typical one would be I step out of the room for a moment, I'll be right back, and then when I come back to the room everybody is gone. Wait a minute, what's with this? And the answer is it's typical dream content. Of course people disappear when you stop paying attention to them because they are not really there.'

Another technique is known as 'mnemonic induction'. Here you repeat to yourself before you go to sleep that you will notice that you are dreaming, and you will achieve lucidity in your dreams. This requires quite a bit of practice.

What are the benefits of lucid dreaming?
LaBerge lists four major ones.

'The first thing is that people definitely like lucid dreaming, they find it a rewarding experience. It's an unusual condition. I'm having this amazing control where I can do things that I didn't think were possible. I can fly, for example.'

Paul Davies says lucid dreams are a liberating experience for a theoretical physicist. He says in a sense you are suspending or manipulating the laws of physics, and you don't have to apply for a research grant!

'The second general area,' LaBerge continues, 'might be simulation, or using the dream state to practise, and that ranges from things like athletic performance, musical performance, social interactions. People have described overcoming shyness, using it as a means of cognitive behaviour modification, overcoming fears. When you are lucid it still feels real, even though you know it's not.' Someone with performance anxiety who is going to play the violin in front of a big audience, for instance, can practise in a dream and have control over how safe they feel. This can relieve their anxiety when it comes to the real thing.

'Then a third area is creativity, enhancing the possibilities of new ideas. For example, artists looking for a new painting would go in their lucid dream to open a door in the expectation that on the other side of this door will be a gallery showing new art. And indeed they open the door and there are these new paintings, and then they remember and reproduce them when they wake up. People have described using that as a means of getting new musical ideas, new ideas in computer programming and relationship management, all kinds of things that are basically using a creative synthesis of our abilities at night.'

Fourthly, LaBerge says, people can use this ability for emotional healing and discovery. 'Dreams have long had the reputation for being the way that people work through problems, get to the point of being able to let go of something, including for example having experience with an encounter with a dead loved one and being able to experience them in a way that lets you actually say goodbye and let go. There are so many different applications of healing in terms of the mental health level of overcoming nightmares, of facing your nightmares and working through them in a way that gives you a sense of empowerment that you can handle these fears within yourself.'

He adds that 'lucid dreaming can give you an opportunity to have an encounter between the unconscious and conscious mind in the dream world that is difficult to arrive at in other places'. He knows this from personal experience: 'As I had a lot of lucid dreams I started to find that they had a personal value. They were meaningful ... So it became an important inner therapy for me, as a means of a personal exploration and discovery of what I was beyond the self.'

A word of caution

Those with certain mental health problems are recommended not to pursue this technique. One example is **schizophrenia:*** sufferers may have trouble distinguishing between dream images, hallucinations and their waking reality. Some scientists are also concerned that lucid dreaming may lead to a psychological confusion between being asleep and awake, which could have a detrimental effect on vulnerable people, such as those with **dissociation** (which I'll look at in Chapter 18).

With those cautions in mind, LaBerge feels strongly that we still need to work on making lucid dreaming accessible to a wider range of people. 'Now you can get much more information out of what's going on in the brain,' he says. 'And there has been a growth of research in the past 15 years in consciousness. So I'm looking forward to more trained neuroscientists taking on this kind of study, because I think it has great potential.' In a nutshell, 'lucid dreaming provides that extra perspective that gives a much broader view of the possibilities of life'.

If lucid dreaming doesn't work for you, there's another kind of dreaming sweeping the world that *everyone* can do – *together.*

Social dreaming

Who can forget the moving scenes shown on TV and social media in early 2020, after areas of Italy were locked down due to the COVID-19 pandemic? People were leaning out of their apartment windows – alone, but creating a type of collective hope and magic by singing together.

In the same way, we dream as individuals, but by sharing our dreams with others we can gain insight into the hopes and anxieties of society as a whole. This is the idea behind the **social dreaming*** movement.

Australian researcher **Dr Susan Long** is co-editor (with Dr Julian Manley) of the book *Social Dreaming*. The concept of social dreaming was developed by the late Scottish sociologist Dr Gordon Lawrence in the early 1980s, when Long was working with him.

She explained to me that in social dreaming people come together, either in person or online, to share their dreams. This is called a **social dreaming matrix**. The process in the matrix is that you contribute your dreams, but nobody interprets anything. Then you start to look at how one dream connects to another, exploring their collective meaning and what it says about the broader social consciousness. The dreams are not treated as if they belong to the *individual* – as in Freudian dream analysis – but as if they belong to the *group*. When people talk through their dreams in this context, they are giving voice to something bigger than their personal ego.

In 2019, the Social Dreaming International Network was formed to share responses to significant world events and widespread crises. In light of the COVID-19 pandemic, this turned out to be very timely. In March 2020, the network held a series of international matrices on Zoom, in response to the emergence of the pandemic.

Franca Fubini, a psychotherapist and organisational consultant, based in northern Italy, is the network's chair. She told me about some of the dreams that were shared early in the pandemic, before it really hit Europe: 'It was as though they were announcing that the big major catastrophe was going to hit … there was an asteroid which was going to fall on Earth and create a big explosion. There were images that talked about something unexpected falling from the sky that would hit the Earth and there would be some kind of disaster. And there were dreams about losing control, like you are driving your car and you find yourself in the back seat and you're not driving any more.

'Many people were having nightmares, including images of asteroids hitting the Earth, vehicles falling from the sky and people jumping out of skyscrapers. Later as COVID-19 spread and saw Italy go into lockdown, more dreams of being out of control were reported. They featured alien attacks, people dying, sickness, distortion and people wanting to run away and hide. Many of the dreams were about **anxiety*** and fear – and few about hope or a bright future.'

Fubini was based in Turin at the time, and people there were finding the lockdown very tough. From her experience in her clinical practice, the collective **trauma*** of the pandemic was reawakening past individual traumas for many people. The social dreaming matrices happening on Zoom gave people the opportunity to access the narrative of the dreams and a key to understanding what was happening in society.

Fubini, Long and other organisational consultants around the world use social dreaming in a variety of contexts: workplaces, conferences, art centres, schools, universities and more. This can help groups to overcome blockages and challenges, by encouraging new ideas and perspectives that emerge from participants sharing dreams and associations.

Long sees huge potential for social dreaming. 'You know, there are some people that say dreams don't have meanings, they're just your brain going haywire, which is absolutely crazy.' Long believes we can usually work out that a dream is giving us a message. She says, 'Psychoanalysis has discovered, and I follow that too, that the messages in dreams are important for our own understanding of ourselves and our own psychological health.

'But if we can also find the *social* meaning in dreams, that is the way in which dreams express a common experience, common fears and hopes, collective memories and collective anticipations, once we can start to understand them, we can then help to understand our collective mental health and what we have to do together.'

Daydreaming

Of course, dreams don't just happen when we're asleep. And research is discovering that daydreams can give us just as many insights and tell us just as much about ourselves as the dreams we have at night.

If you often find yourself gazing out the window, in a world of your own thoughts, your mind wandering from one random thing to the next, you're not alone. It's estimated that we all spend half of our waking lives daydreaming or mind-wandering. If you're trying to focus on what you're doing, it can feel unhelpful, but it's actually an essential part of human life, and something that sets us apart from other animals.

Daydreaming is the passion of **Dr Muireann Irish**, now Professor of Cognitive Neuroscience at the University of Sydney. She says there's lots of activity going on in your brain when you 'zone out'.

Neuroimaging studies show that when healthy individuals daydream while their brain is being scanned, the brain is actually hard at work. Scientists see an increase in activation of the brain's **default mode network**,* which I looked at in Chapter 4 (Creativity). It's engaged when we reminisce, imagine and engage in self-reflection and creative thinking.

Daydreaming and creativity

Often, when we're trying to work out a difficult problem, we get stuck. It seems that if we put the idea aside for a while and let our mind wander, free from work demands and the constant influx of information, sometimes a creative breakthrough happens. (It's a bit like the 'Aha!' moment I looked at in Chapter 4, and the creative breakthroughs we have after 'sleeping on it', which I examined in Chapter 5.)

'And so some researchers have suggested that we need these periods of incubation where we are really not focusing on the problem per se,' says Irish, 'but in the background our daydreaming is facilitating us to bring different ideas together and to make associations that we may not necessarily have made before. And then we have this creative breakthrough.'

Comedian Woody Allen attributes some of his best ideas to tuning out in the shower, and author JK Rowling says she came up with her idea for the Harry Potter novels while daydreaming on a long train journey.

However, Irish warns: 'There is the other side of the coin that if we daydream too much we are not being attentive to the present moment and we are not focusing our attention on tasks that do need it. So I think it's a question of balance.'

Mental health and daydreaming

On the downside, daydreaming can contribute to **depression*** and **anxiety,*** by exacerbating the tendency to ruminate on negative thoughts. In depression, people can get caught in a process of thinking negatively about their *past*. With anxiety, it can be problematic in the opposite way: people constantly envisage the *future* negatively and worry about that. In these cases, some of the techniques used in **mindfulness*** could be helpful, by assisting people to rein in their negative thoughts and come back to the present moment.

These days, though, it seems to me we often limit our opportunities to daydream in a *constructive* way. Many of us are spending less time simply gazing into space, allowing our thoughts to drift. Instead, we are hooked into our digital devices at every opportunity. This prevents us from regularly letting our imaginations loose and coming up with new and unexpected ideas and associations. As Irish says, 'I do think when you see how frequently we're checking our emails on the bus on the way to work, we are not really taking the time out to let our minds wander or to doodle or to draw pictures any more.'

She says that this is particularly an issue when it comes to children. It's known that imaginative play can help them develop their social skills, be more creative and foster their sense of self-identity. We need to be mindful of how we can encourage children to use their capacity for daydreaming and imaginative play, particularly in schools, without compromising the need for focused attention on lessons such as maths and reading. Again, it's a matter of finding a healthy balance.

'I think that these higher-order forms of thought enable us to bridge the gap between mind and brain,' Irish says. 'It seems that we have evolved to have these fanciful forms of thinking to escape from the present moment, and so the question arises as to why, and what purposes these forms of thinking serve.'

Part II

NEW INSIGHTS INTO
LIVING WITH
MENTAL 'DIFFERENCE'

7

AUTISM AND 'NEURODIVERSITY'

Conversations with:

John Elder Robison, autistic author and autism advocate

Nicole Rinehart, Director, Krongold Clinic, Faculty of Education, Monash University, Melbourne

Steve Silberman, science writer

Thomas Kuzma, Online Social Coordinator for Student Affairs and Events, Landmark College, Putney, Vermont

Tim Sharp, autistic artist

Judy Sharp, Tim's mum

Jill Bennett, Scientia Professor of Experimental Arts and Director, Big Anxiety Research Centre, University of New South Wales

Tom Middleditch, Autistic Artistic Director, A_tistic theatre company

Hannah Belcher, autistic lecturer in autism and mental health research

Francesca Happé, Professor of Cognitive Neuroscience, King's College London

'The weird guy': John's story

'I was the unwanted child,' says John. 'I was the one they didn't want to play with and the one they didn't want to be friends with. And that was something that was very, very painful, being isolated.'

John Elder Robison has **autism spectrum disorder (ASD).***
People with this condition often have difficulty interacting and
reading the emotions of others – so I didn't expect my conversation
with John to so deeply move me.

He was born in 1957, when there was a very poor understanding
of ASD. 'So when I didn't say or do the right things or I didn't
respond appropriately, people just thought I was a bad kid – I was
selfish, I was self-centred, I was in my own world, I was lazy,
I was stupid – and they kind of discarded me.'

It also affected his schoolwork. 'I was not able to do the things
people asked of me in school, so they would give me failing grades
even as I thought I could do the work, and it made me very angry,
and eventually I stopped going to school. The last grade that I passed
was the ninth, and I left home and I joined a rock 'n' roll band, and
that was my ticket to independence, and that came about because I
had this fascination with music and electronics and I learnt how to
fix and then build sound equipment for musicians.'

But he didn't appreciate music in the same way as other people.
'What I appreciated was the beauty of delivering whatever the
musicians gave me true to form. So *you* might think he sings
so wonderfully as a tenor or alto or soprano, and he sings these
passages with such nuance and emotion, but I would never say
anything like that. I would think that I was very successful if
I delivered his voice to you clearly and crisply and without any
distortion. To me *that* was the beauty of it.'

By the mid-1970s, John was an engineer working for sound
company Britannia Row Productions, putting together sound
systems for many British bands touring North America and
Australia. 'And I had equipment out with soul bands, disco bands,
progressive rock, all kinds of different music. And again, it wasn't
that I liked one kind of music or another, I just thought it was really
cool if I could build a sound system that could deliver country music
and then do heavy metal with the same stuff and do both well.' He
became known as 'the weird guy' who talked to machines and built
amplifiers – and the nickname on his photo ID was 'Ampy'.

John reflects: 'For me the pain of isolation, it hurt me a lot, but being isolated is what gave me time to concentrate on music. And then my autistic focus, my other autistic traits, allowed me to become a star engineer. So that pain was kind of a part of what made me a success.'

He worked with rock bands like Pink Floyd and Kiss, a band renowned for their crazy costumes and wild live performances. John was the guy who made Kiss band member Ace Frehley's guitar smoke: a bit of electronic mastery that he's very proud of!

John describes it as a wild world of drugs, liquor and guns. He tried to stay clear of that scene as much as he could. 'I didn't understand why I was the way I was, but I understood that I had trouble reading people and I had trouble figuring out what they really meant when they said something. And being drunk or high on coke just made me a drunk or a high fool, and so I wanted to stay clear of that because I had enough challenge without it.'

He decided to seek an engineering job with a less crazy lifestyle, and took a job with an electronic toy and game company. He got married when he was 25.

'I went through four jobs in succession, never really fitting into corporate culture. I decided I'd be better off on my own, and I didn't have any money but I always loved tinkering with cars. So I started a business fixing them and it was a success.'

His son Jack was born shortly after his thirty-third birthday.

In his auto-repair business, one of his customers was a therapist, 'and he watched me, I guess, and talked to me for a number of years'. When John was around 40, this man helped him realise that he was on the autism spectrum. The man told John: '"I thought a long time about whether I should say anything to you about this because you are successful, you've got a wife and a kid and business, but I think that knowing why you are like you are could really change your life." And boy, it really did. It was stunning for me to hear that.'

He adds: 'For the first time in my life I had a non-judgmental, non-negative explanation. No, I wasn't stupid, retarded, or defective ... And that was really liberating. It's a big, big deal'

So, 'I thought to myself, "Well, I'm going to, by God, make myself act like everyone else." And today, autistic rights advocates would say you shouldn't have to do that, the world should accommodate you. But you know, we can only go so far in asking the world to accommodate us. The fact is, if our inability to read your body language or your facial expressions causes you to think I'm a callous jerk in the first 10 seconds you meet me, you are never going to know me long enough to decide I'm a nice guy.

'So it's incumbent upon me to know how to act when I meet you. And knowing the ways in which autistic people like me were different, I was able to change my behaviour, and it was really a magical transformation. I began to have friends for the first time in my life.'

John decided to take on the mission of raising awareness about autism by writing several books – one of them a *New York Times* bestseller – and speaking widely about his experience.

'I think that it has become really a passion of mine to go out and speak for the rights of autistic people,' he says, 'to show the rest of the world that we autistics, we are not smarter, we're not better, we're just different, and the rest of the world needs different people.

'We are a part of diversity, and we are something that's needed. And I see my mission is to go out and spread that word. I think it's an incredible honour, and I would say that we autistics are just coming into our time. We have to stand up and demand the acceptance other groups have achieved.'

(This isn't the end of John's story. More recently, a revolutionary new brain therapy called **transcranial magnetic stimulation (TMS)*** has given him a whole new way of seeing the world. I'll pick up the tale again in Chapter 19, when I look at TMS in detail.)

Autism: A neurodiverse way of being

Neurodiversity* is challenging our notions of what's mentally normal. It's a revolutionary cultural movement grounded in the idea that we are all born different, whether we fit into what's traditionally been considered 'the norm', or whether we deviate from that 'norm'. Each person is unique in their way of being and their way of perceiving the world, and that uniqueness is something to celebrate.

The term 'neurodiversity' was coined by Australian sociologist Judy Singer in the late 1990s. Her mother and daughter both lived with autism, and she felt she had autistic traits herself. She was struck by the attitudes many people had towards those with autism: the kinds of attitudes John Elder Robison has experienced for much of his life.

Neurodiversity has now extended *beyond* autism to include a whole range of conditions and different ways of being. Those who have these conditions are considered 'neurodiverse'; those who do not are regarded as 'neurotypical'. I'll introduce you to some of these other ways of being later in Part II.

To some extent, however, the concept of neurodiversity has been divisive, both in the autism community and among autism researchers, with some arguing that it may only exacerbate the 'us versus them' mentality that many people with autism already feel. Others believe it's only relevant to people with *milder* forms of the condition and neglects those who struggle with *severe* autism.

Regardless of the controversy, the idea of neurodiversity has taken off in most parts of the autism community, and has become what Singer hoped: a way of giving autistic people a sense of being in control of their own lives and stories, rather than being perpetually defined as inferior by the medical establishment and society as a whole.

As John Elder Robison puts it: 'I believe with all my heart that autistic people deserve a place ... really a place in the sun, it's our time.'

So, what exactly is autism?

Understanding of autism spectrum disorder has grown hugely in recent years, but it remains one of our most controversial health issues.

An autism diagnosis is based on the observation of certain behaviours, but there is no reliable biological test. It's believed to be four times more common in boys than girls – something I'll look into later in the chapter.

Autism is characterised by difficulty with communication and social interactions, restricted interests, repetitive behaviours and sensory sensitivity. It's described as a spectrum because the symptoms and severity vary vastly between individuals and can change over the course of a lifetime. As US autism researcher Dr Stephen Shore suggests, 'If you've met one person with autism, you've met one person with autism.'

The exact cause of autism is not known, but the consensus is that it's a combination of developmental, genetic and environmental factors. (What is certain, though, is that autism is *not* caused by bad parenting or childhood vaccinations – but more on that later.)

Estimates of worldwide prevalence vary quite widely, but Autism Spectrum Australia (Aspect) recently revised its rates from 1 in 100 to 1 in 70 Australians. This figure reflects new national and international research, as well as recent changes in how autism is diagnosed.

The fifth edition of the American Psychiatric Association's (ASA's) *Diagnostic and Statistical Manual of Mental Disorders*, or **DSM-5*** – often referred to as 'the psychiatrists' bible' – lists three levels of autism severity. They range from a mild awkwardness in social situations, to a very limited ability to speak and engage in social interaction. DSM-5 was published in 2013, and in March 2022 the ASA published a text revision, DSM-5-TR, which contained further small changes to the definition of autism.

Perhaps the biggest change introduced in 2013 was eliminating a separate category for **Asperger's syndrome.*** According to **Nicole Rinehart**, Director of the Krongold Clinic at Monash

University, people with Asperger's 'are normally intelligent, have very good verbal skills but still have difficulties in the areas of social communication and stereotyped, repetitive behaviour patterns'. DSM-5 now classifies this as being at the milder end of the autism spectrum. Many people diagnosed with Asperger's before 2013 are still perceived as having Asperger's syndrome – and some still consider it an important part of their identity. (This is in fact the form of autism John Elder Robison has.)

So why has the current diagnosis of autism taken so long to emerge? Why is it still so controversial? The answers lie in a disturbing story involving two rival researchers that stretches back 80 years, and has led to huge misunderstandings that persist to this day.

The dark history of autism

In the mid-1990s, US writer **Steve Silberman** noticed a high proportion of Silicon Valley workers who had children with autism – and a massive increase in autism diagnoses. He decided to investigate. What he uncovered was a complex history – and his work became part of a huge change in our understanding of autism spectrum disorder that is still in progress today.

I spoke to Silberman in 2015 after he'd released his book *NeuroTribes: The Legacy of Autism and the Future of Neurodiversity*, which later became a bestseller. The story that caused him to rethink our understanding of autism was a conflict between two psychiatrists in the 1940s.

This is the way the story of autism was told in the past, Silberman says: the condition was identified in 1943 by a child psychiatrist in Baltimore named Leo Kanner, who wrote a landmark paper on 11 patients. A year later, an Austrian doctor named Hans Asperger also described autistic children, in a paper that was considered a minor contribution compared with Kanner's work.

After several years of research, Silberman found that this timeline was completely incorrect.

'What I discovered is that Hans Asperger and his colleagues at the University of Vienna in the mid-1930s discovered what we now call the autism spectrum, which is a condition that lasts from birth to death, that encompasses everyone from children who are unable to speak and will never be able to live independently, all the way up to chatty adults who become assistant professors of astronomy etc. So Asperger and his colleagues had a very, very broad, and I have to say, humane conception of autism as a spectrum or a continuum.

'Leo Kanner, by contrast, had framed autism as a very, very rare form of childhood **psychosis**,* and in fact he would eventually blame it on parents, so-called refrigerator mothers and cold and unaffectionate fathers, and that had a disastrous effect, because a couple of generations of autistic kids were put in institutions and more or less forgotten and became invisible to the rest of society. So it seemed to be really rare because autistic people were sort of shut away behind the walls of these institutions.' Silberman adds: 'It cast a terrible shadow over autistic people and their families for most of the 20th century.'

Kanner and Asperger crossed paths and worked together to some extent, but Kanner failed to acknowledge Asperger's work, effectively burying the concept of autism as a spectrum – that was, until 1968, when British psychiatrist Lorna Wing studied the prevalence of autism and found many more undiagnosed children than Kanner's narrow model would have predicted. Later, she came across a reference to a paper by Asperger and discovered that the concept of autism was much broader than previously thought. In 1979 Lorna Wing and colleague Judith Gould published a landmark study of children in Camberwell, South London, which was one of the first epidemiological studies of autism, reporting a prevalence rate of 1 to 2 per 1000 people. In her 1981 paper, Wing suggested the term Asperger syndrome, which led to this clinical diagnosis being included in the American Psychiatric Association's fourth edition of the *Diagnostic and Statistical Manual of Mental Disorders* in 1994.

'And so she sort of quietly swapped out Kanner's model from

diagnostic guides like the DSM, the so-called bible of psychiatry, and swapped in Asperger's spectrum model, and that was when the so-called tsunami of autism began,' Silberman said.

Over the next two decades, Wing popularised the spectrum model of autism, and argued that Asperger's syndrome was a mild form of autism. What followed was a huge rise in the number of diagnoses.

With this came a long list of apparent explanations for this rise – the most notorious being from British medical researcher Andrew Wakefield, who attributed the spike to childhood measles, mumps and rubella vaccinations. His work has since been completely discredited, but has remained at the centre of heated political and cultural debate.

'Well, here's the thing, many parents believe that Andrew Wakefield was a martyr in a sense and is being persecuted by the medical establishment and the pharmaceutical industry,' Silberman explains. 'But the problem is it's not true, it's not a true story, that's not why the diagnoses kept rising. And so it makes parents terrified about the wrong things, like terrified about inoculating their child against measles.'

He adds: 'The other thing is that it has had a biased effect on the direction of research, in that we keep pretending that autism is this modern anomaly, that it's like the unique disorder of our disordered world. So even if it's not vaccines, people say, well, it's pesticides or it's Wi-Fi or it's video games or it's antidepressants in the water supply, or it's ... something is different. Actually, what's different are the diagnostic criteria for autism. And now finally more autistic people and their families can get diagnoses.'

Silberman considers the new movement of **neurodiversity** central to reframing society's view.

Standing up for the little guy: Thomas's story

I spoke to **Thomas Kuzma** in 2013, just as the DSM-5 was about to be released – and about to do away with the separate diagnostic criteria for Asperger's syndrome. He was diagnosed with Asperger's when he was 16.

'This is why I always mention the dark years as my high school years,' Thomas told me. 'It always felt like there was nothing, nowhere, everything was darkness. Asperger's to me felt like a light being shone upon me, like, I don't know, some kind of big beam coming down saying this is what makes you, you. And I finally found something out about me that I could be proud of.

'At the same time I was watching a show called *Boston Legal* which had an Asperger's character in it, and I watched this one episode where I saw him fighting for the little guy, and that's what I told myself: "This is what I'm going to be, I'm going to be Thomas Kuzma, the guy who defends the little Aspie guys and helps them succeed in life." So I go round talking about Asperger's syndrome.'

Since his diagnosis, he's learnt more social skills. 'I have found out many things about people, I've learnt things like body language and facial expressions. I have learnt things like empathy through friends and through what I have seen through other people. So when it comes to relationships I have done my best to think about the other person, not just think about myself.'

But it's been a struggle at times. From some challenging dating experiences, he's learnt not to make too much or too little eye contact, and not to keep on talking about the same thing if the other person is not responding well.

Going for job interviews has been tough too. 'Now, I'm a person who presents really well for someone who has Asperger's syndrome, and when people see that, they think, "Oh, this guy, he's just making up excuses, he doesn't really want to work", and I'm like, "No, I would like to work in this field, I have the skills and I've got the experience as well." And unfortunately many times I just haven't received the job itself.' It can also be hard when he *does* get the job: 'When it comes to coming to a new place, that tends to muck it up a little bit, and because I'm trying to get used to something new I usually need someone who is there for when I start.'

As with many people on the autism spectrum, Thomas can get sensory overload at times. He told me about a time when he had a filming job, and they were shooting a pub crawl. The summer heat was intense, and many people were drunk, pushing and pulling him, wanting to be filmed. He said the experience broke him. He couldn't work for two months, or even go outside.

Thomas has developed certain strategies to help himself when he's overwhelmed and is about to have a meltdown. One of his favourites is finding somewhere quiet, taking a cold bottle of water and his iPad and listening to some smooth jazz.

Thomas's experience highlights some of the often-underestimated difficulties of having a milder form of autism – or Asperger's syndrome, as it was previously known. Their struggles are not as obvious or easily acknowledged as those of people with more severe autism.

But for Thomas there's also a positive side to being on the spectrum: 'Back when we first found out, Dad was a little bit worried, but as time has gone on he has become more welcoming. He's found out it's not a disability but more like a gift, because every Aspergian, every person with all kinds of mental conditions, they've got gifts that should be celebrated in the world.'

Thomas notes now that even though he used the terms Aspie and Asperger's syndrome earlier in his life, now he no longer uses those terms and strongly advocates as both autistic and neurodivergent.

He now works for a Neurodivergent College in Putney, Vermont, known as Landmark College. He is their Online Social Coordinator for Student Affairs and Events, helping neurodivergent college students mingle and learn about social skills.

When it comes to autism-related projects, Thomas's mantra is 'nothing about us without us'.

Laser Beak Man: Tim's story

The day **Judy Sharp**'s first child, **Tim**, was born was the happiest of her life. 'By a year Tim was walking and he was hitting his

milestones,' says Judy, 'but there were other things happening
that were a little alarming. One of them was a great sensitivity
to sound or different sensations, and I had no doubt that Tim was
intelligent and had a lot of abilities, but there were a lot of unusual
things happening and I'd never seen them in my life before.

'But he was always the love of my life, he was the joy of
my life, but I felt I had to do something to be helping him to
get through these everyday things that were really upsetting
for him.'

When Tim was three years old Judy took him to a specialist
doctor, who diagnosed him with autism. She was devastated by
the doctor's predictions that Tim's difficulties would be so severe
that he would never speak, go to school or have any quality of
life, and that she should give up on him.

Understandably, Judy rejected the prognosis, and she placed
him in intensive therapy to try to overcome the things that were
so terrifying for him.

In the early days, Judy's greatest fear was that because Tim
couldn't speak, he could never tell her what was wrong, and so
she'd never be able to fix things for him. She became more and
more frustrated and desperate. When Tim was four she sat down
and drew for Tim. A year later Tim took up a pencil and began
drawing for himself. 'And it was clear from the first drawing
that he had something,' says Judy. When he was 11, driven by
his passion for superheroes, Tim created Laser Beak Man. He's a
brightly coloured, quirky character drawn boldly in crayons, pens
and pencils. He dons the classic superhero cape with a large letter
'L' displayed prominently on his chest.

'He has a laser beak in his face,' Tim explains; 'he doesn't use
it much. He's been in art galleries, TV shows and a play as well.'

'Tim is the first person in the world with autism to have his art
turned into an animated TV series,' Judy told me. 'We've been to
New York to work on the development of the stage production of
Laser Beak Man. Tim has become a keynote speaker and was in the
final three for best speaker in Australia. He's had a book. We've

travelled the world, there was a music festival started in his honour for autism awareness in Nashville, Tennessee. His exhibitions sell out, he has shown art at the Metropolitan Museum of Art in New York City. Some landmarks here in Brisbane have lit up with his art. It just blows my mind, I can hardly believe it.'

Judy attributes some of the success of his art to the fact it is so original and so uniquely Tim. His drawings often reflect the very literal interpretation of language that comes with autism, and they also have a strong element of wicked humour and irony.

One example she gives was inspired by the experiences they've had together in cafés. Tim would hear people saying, 'I'll have a flat white ...', 'I'll have a flat white ...'

So, Tim drew a woman and a waiter outside a café. In the middle of the road in front of them is a white man lying completely flattened, and Laser Beak Man is driving a steamroller away. So that's a flat white!

Another cartoon is called 'Picking up Chicks' – which depicts Laser Beak Man in a farmyard, picking up fluffy yellow chickens. Judy chuckles about this one, saying that a lot of men like it, wishing that dating were that easy for them.

Tim Sharp is now in his 30s and his successful art career has made a world of difference to his confidence and his happiness. I asked Judy how important she thought Tim's art was to his sense of wellbeing.

'I think it's very important, actually, because it is that connection with the community, and it gives him a great opportunity to meet with a lot of people. And there's an expectation for Tim with all these opportunities that he can do it, and that's always encouraging, when somebody believes in you.'

Attitudes to autism have changed a lot in those 30 years too. 'I think a lot more people are aware of autism now,' says Judy. 'Everybody seems to have heard the word. I'm not sure that there is a great deal of understanding of what it is. I think there's definitely a lot more opportunity for people with autism to live a really fulfilling life.

'I've seen that going forward, but at the same time there are also some really huge gaps that haven't caught up for people who require a lot of support, there's still not that recognition of the part that autism plays in their life. So we are doing some things really well and some things are just taking a bit longer to catch up.

'But getting the information out there is very important, because once people know that we are more the same than we are different, we can break down a few more barriers and have more inclusion.'

The Big Anxiety

I'm fitted with a virtual reality headset and led through a door, to see stark white walls straight ahead of me. When I turn and look behind me, there's no longer a door, but two windows. I look out of them and see nothing but pitch black.

Another door appears in front of me, and opens to reveal the most beautiful mountain scenery. Gorgeous bright sun lights up a rich expanse of green grass against a clear blue sky featuring puffs of white cloud.

The experience is quite powerful. Immediately I feel calmer and more positive.

When I look back into the room, and through the two windows, I can see a bright green and sunny landscape of rolling hills dotted with yellow daisies through one, and through the other misty mountains with the sun just trickling through the clouds. At my feet, what were once floorboards are now grass and pebbles.

I turn around and open another window in front of me, and suddenly all the walls disappear. The window and door frames are floating in space, leaving me in the middle of a beautiful sunlit landscape, with light snowflakes flitting down, collecting on the leaves of the trees against a background of sweet birdsong. It's a truly uplifting experience.

I'm at the second The Big Anxiety festival, held in Sydney in late 2019. The Big Anxiety brings together artists, scientists and communities to celebrate mental diversity and explore the experiences of **trauma*** and suicide survivors in a way that promotes empathy rather than fear and stigma.

The Edge of the Present, the powerful virtual reality installation I've just experienced, grew out of suicide prevention work and was developed by artists Alex and Michaela Davies in collaboration with scientists and survivors. Its focus is on learning to cultivate positive future thinking.

Alex explains to me that the room I was in represents my inner world. It begins as quite negative and is transformed over time, depending on my level of engagement with the outside world.

Michaela, who's also a psychologist, says it's about helping people stop feeling that they are trapped in a particular mode of thinking. It's not about saying everything is amazing, but about being open to the possibility of change. The more you explore, the more open you are to the entry of the external world into your internal world.

Jill Bennett, festival founder and Director of the National Institute of Experimental Arts at the University of New South Wales, believes we need to start thinking about neurodiversity in the same way we think about *cultural* diversity: as something to be celebrated. She says there is no better forum for doing this than the arts.

'Empathy is a really important concept if we are trying to sustain any kind of diversity. There is actually some evidence in neuroscience that the mechanisms associated with empathy, like sharing emotional and motivational states, are limited to people we like, or what we call in-groups. And what that means of course is if it's an adaptation that serves to reinforce social bonds, what is actually going on is we are directing empathy *away* from other groups that we *don't* identify with. And I think we perhaps need to think of empathy as something more encompassing than that.'

Bennett is passionate about the concept of neurodiversity. She thinks society is now inclusive and empathetic to a point, but still worried about the idea that autism and other different ways of being are 'disorders'. 'So I think the best comparison is with those other struggles for equality and parity around gender and race and ethnicity. We have to make the same steps with neurodiversity, and I'm sure we will.'

Autism in the theatre

The 2019 Big Anxiety festival wasn't the first place where I met Jill Bennett. The year before, she was a speaker at a gathering I attended hosted by the Arts Centre Melbourne. The play *The Curious Incident of the Dog in the Night-Time*, written by Simon Stephens and based on the well-known novel by Mark Haddon, was being performed in Australia for the first time. It's the story of a mathematically gifted 15-year-old boy with Asperger's syndrome who sets out to solve the mystery of who killed a neighbourhood dog. The book and play both highlight the importance of nurturing empathy for the way people on the autism spectrum perceive the world.

The Arts Centre Melbourne saw the play's Australian launch as an opportunity to investigate some of the issues involved in exploring neurodiversity within the arts and the theatre, through an event called The Deep Dive.

Another presenter at the event, **Tom Middleditch**, introduced himself to me as a philosopher, playwright and Artistic Director of A_tistic, a theatre company that employs people who are neurodiverse. Middleditch himself is on the autism spectrum. He told me he only wears maroon, he's very sarcastic and makes pitch-black jokes in a straight voice, and if no one gets the joke, then that's a joke for him to enjoy. He thinks that most people would find his communication style 'passable'. Middleditch was also diagnosed with ADHD when he was about six, and the condition gives him seemingly boundless energy.

He's always been interested in theatre, and in 2015 he began A_tistic in response to what he thought was a lack of appropriate representation of the autistic experience within the arts. For the company's first production, called *Them Aspies*, they created a model to train non-autistic actors to play autistic characters authentically, and give them empathy for those on the spectrum.

As Middleditch says, 'Theatre has always been about drama and it has always been about people negotiating.' So it's the ideal forum for expanding people's minds and acceptance of the neurodiverse world.

Taking off the mask: Hannah's story

Englishwoman **Hannah Belcher** was diagnosed with Asperger's syndrome in her early 20s. She'd been seeing therapists throughout her adolescence, but the diagnosis came as a bit of a shock.

'I think the main problem was that my symptoms weren't as obvious as they are in males. So things like eye contact I was perfectly OK with ... And I think doctors just saw my **anxiety*** and just saw that as the main diagnosis without really considering why is that there. I think they saw *that* as the challenging behaviour problem and not what was underlying it.'

Social interaction became so challenging for Hannah that she dropped out of school at the age of 14. Other autistic traits she experienced were extreme sensitivity to noise, colour and smells and obsessional musical interest, where she would just listen to the same song over and over again. Her relationship with food when she was a child was very particular too: she would have to have her food cut up and presented on the plate in a specific way and would only eat certain items continually until she got sick of them – then she'd move on to the next thing. She told me that she's still a bit like that now, but slightly better.

Belcher thinks girls and women with autism mask their symptoms because of the social pressure on them as they are growing up. They're taught to be polite and adaptable, and they become very good at what she calls social mimicking. She became so passionate about the topic of female autism that she's completed a thesis on it and now lectures at the Institute of Psychology, Psychiatry and Neuroscience at King's College London. Her book *Taking Off the Mask* is about how autistic people can stop 'camouflaging' their behaviour and learn to be themselves.

Many people tend to think of autism as a male disorder, but research is now showing that Hannah's experience is common. Girls have different symptoms that often cause them to slip through the net – misdiagnosed or undiagnosed by clinicians.

Why do girls go undiagnosed?

Francesca Happé is currently Professor of Cognitive Neuroscience at King's College London. Most recently her work has focused on mental health in autism, particularly in less researched groups such as women and the elderly.

One of the best-known women with autism, she says, is Temple Grandin, Professor of Animal Sciences at Colorado State University, who has written extensively about her experiences as an autistic person: 'She is an extraordinarily brilliant woman who has several degrees and a PhD and designs livestock facilities.'

'How is she representative of women with autism? Well, she's highly, highly intelligent, she talks about her ability to compensate and to cope, and she also interestingly talks about having made great progress even later in life.'

About four times as many males as females get an autism diagnosis, Happé says, but she adds: 'I think we know very, very little really about how autism presents in girls and women. There are *some* studies, but the main problem is that the studies start in a clinic. And so you can see there is a circularity. If we are missing women and girls with autism because we are not good at recognising them, then studying those we *do* spot isn't going to tell us very much about the ones we miss.'

Because ASD can be missed in girls, they often don't discover it till later in their lives – which is what happened to Hannah.

One of the reasons ASD in females may be missed by clinicians, Happé says, is that it may look different from ASD in males. As Hannah's research suggests, girls may be better at covering up the more obvious characteristics of autism.

She agrees with Hannah's theory about social mimicking. 'Some women with autism describe a strategy of copying somebody. They pick a female in their class or their workplace and they just copy everything about that person – how they dress, how they act, how they talk – and that kind of masking strategy we don't see very much in boys and men with autism.'

Typically a clinician assesses a child for autism traits by looking for rigid, repetitive behaviour and special interests. A boy with autism might have a fascination with electricity pylons and know all the facts about them, which will alert the clinician to the possibility of autism.

On the other hand, a girl with autism may be fascinated by a particular pop group and learn all the facts about them, and the clinician is likely to think that's fairly normal. 'So unless he digs deeper and finds out actually she has no interest in going to hear them perform or even listening to their music, she just collects the facts, then otherwise the clinician is going to be fooled into thinking, "OK, this isn't autism."'

Another reason why autism may be missed in girls and women, Happé suggests, is what's referred to as diagnostic overshadowing, where a girl with autism is diagnosed as having an eating disorder, for example, and no further questions are asked, so her autistic traits are not picked up.

There's also growing suspicion that the underlying biology of ASD in males and females may be different. The 'female protective effect' is a theory that suggests girls with autism may have a larger genetic dose than boys, which means that while girls may be generally less susceptible, those who do have autism may be more significantly affected.

'In general, males are more affected by all neurodevelopmental disorders,' says Happé. 'We don't know why exactly that is. Simon Baron-Cohen [Director of the Autism Research Centre at Cambridge University] has a theory that it's about exposure to testosterone *in utero*, predisposing the male brain to the autistic pattern.'

Baron-Cohen's 'extreme male brain' theory contends that we all have a balance between empathy – our ability to understand *social* systems – and systemising – our capacity to understand *non-social* systems. The male brain, he says, is highly developed in systemising and less so in empathising – and autism can be thought of as a very extreme version of the male brain.

Happé finds Baron-Cohen's theory interesting, but she doesn't think it provides a complete explanation. Clearly there is a lot more work to do.

'I think a lot of researchers believe that if we understood why more males than females are affected with autism we would have a better understanding potentially of either the genetic or environmental influences on autism,' she says – the ultimate goal being 'to make the outcome for people with autism the very best that it can be'.

8

HEARING COLOURS, FEELING OTHER PEOPLE'S PAIN

Synaesthesia

Conversations with:

Eliza Watt, synaesthete

Rachel Watt, Eliza's mum

Professor Anina Rich, Director, Perception in Action Research Centre, Macquarie University

Nina Norden, synaesthete and artist

Mark Ledbury, Power Professor of Art History and Visual Culture, and Director of the Power Institute, Sydney University

Helen Thomson, author and Head of Features, *New Scientist* magazine

Seeing sounds: Eliza's story

What colour is Monday for you? What kind of personality do you associate with the number four? Can you taste the colour purple, and does saxophone music sound orange or blue?

If those questions aren't too strange to you, you may be a synaesthete – like **Eliza Watt**. I met her and her mother, **Rachel Watt**, when Eliza was a confident and thoughtful 12-year-old.

For Eliza, everyone has a coloured aura around them, even if she hasn't met them before. I asked her if she could see an aura around *me*. She said that before we met her mother showed her a picture of me and I was yellow – but she decided I was a mixture of yellow and orange after hearing my voice! If she meets someone whose aura is a colour she doesn't like, it can be awkward, because it makes her cringe, and the person may wonder what's going on.

Eliza didn't think she had a colour herself until one day she looked in the mirror and was surprised to see a light peach and rose-gold aura around herself. Her mum, Rachel, is a pinkish-red, and her dad is blue and sometimes a bit green.

Numbers also have personalities for Eliza: 'Seven is not very nice. He's sort of a bully, and 17 is also a bully, but he doesn't do a lot of the bullying, he just advises. So they're not a very nice pair, but six is like a mothering number, and that's sort of a reddy, pinky colour. And two is really nice and little, but she wants to speak her mind to the bullies, like 7 and 17, but she doesn't feel like she can. She's not brave enough to.'

Eliza plays the saxophone, and one of the things she loves about that is seeing colours come from the back of her head and shoot off really fast to a point about 3 metres away. Overall, the saxophone is a yellow instrument to her, but the sounds are often dark blue or green. When she plays it's more yellowy green and sometimes pink, because she likes to play in the higher octaves and pink is a high colour for her.

Eliza told me that she only found out there was a name for her way of perceiving the world when she was about seven, and she was surprised that not everyone saw things as she did. Meeting her, I was greatly struck by her sophisticated ability to articulate her synaesthetic experience, and the creativity she draws from it.

What is synaesthesia?

Professor Anina Rich is the Director of the Perception in Action Research Centre (PARC) at Macquarie University, where she heads up **synaesthesia*** research. 'So what exactly *is* synaesthesia?' I asked her.

Rich explained that 'people have often described it as a mixing of the senses. In fact it's an umbrella term that can encompass a lot of different types of experiences that all fit within a sense of having an extraordinary response to a very ordinary stimulus.'

It's estimated that between 1 and 3 per cent of the population experience synaesthesia, and it's about twice as common in females. It's not a disorder, more an alternative way in which people experience the world.

It's thought that **grapheme colour synaesthesia** – where letters and numbers evoke colours – is very common, as is **auditory visual synaesthesia**, in which different types of sounds create specific visual experiences. (Eliza Watt seems to experience both of these.)

In one study, Rich and her colleagues looked at experiences evoked by single tones from different musical instruments. They found it was not just colour that was perceived but geometric shapes as well.

They found that synaesthesia builds on mechanisms we all have. For example, if non-synaesthetes hear a low tone played on the piano, then a high tone, and they are shown shapes, they will say a small, light-coloured object goes with the high note, and a big, dark shape goes with the low note. This suggests that these connections are in us all, and just more pronounced in synaesthetes.

'Traditionally people talked about synaesthetes as having hard-wired connections between areas that the rest of us don't,' says Rich, 'so that might be, in auditory visual synaesthesia, the auditory cortex, the first place where you process sound; or you might have a connection to the visual cortex, perhaps, the area that is involved in colour.

'There's now evidence that probably most of us – actually all of us – have connections between our sensory cortices in that way. So it's perhaps not so unusual to have that connection. What might be different is that synaesthetes might have stronger connections than the rest of us.'

One hypothesis is that synaesthesia is on a continuum, with synaesthetes at one end and, at the other, those who have no mental imagery at all, a phenomenon called **aphantasia**.

Help or hindrance?

'It's actually quite infrequent to find somebody who has a problem with their synaesthesia,' says Rich. 'If anything, most people think of it as an unusual gift. And most people think it's just strange that we're even talking about it, because isn't that the way *everybody* sees the world?'

Some people do find it overwhelming, though, when they have several types of synaesthesia at the same time – for instance, strange experiences involving hearing, smell and touch all at once. It can also be a bit disconcerting when there's an incongruence between the synaesthete's perception of something and the reality. 'In everyday life it probably just is a little bit of a "Oh, that doesn't look right", you know, "fire extinguisher" is written in the wrong colour, that sort of thing,' says Rich.

She and her team have tried to get an objective measure of the synaesthesia experience in the lab. For example, they might present the letter 'A' to a synaesthete, first in a congruent colour, or in a colour that they usually synaesthetically perceive the letter 'A' to be (e.g. red), and then in a colour that they don't perceive the letter to be – say blue, which is incongruent for them. The researchers then measure the synaesthete's reaction times in naming the display colour. Synaesthetes tend to be slower to name incongruent colours, which shows the implications these incongruences may have for people in their everyday lives.

Sydney University synaesthesia researcher Joshua Berger became interested in whether grapheme colour synaesthesia – when people associate specific colours with particular numbers – affects learning. One of his fellow students with this condition found that mathematics never made much sense to her because, for example, it was incongruous for her that a blue three times an orange four should equal a green 12. For others in her class, none of the numbers had a colour, so it was more straightforward for them. This led to her maths anxiety.

In response to his friend's difficulty, Berger developed a coloured calculator to see if that could reduce maths anxiety for his friend and others with grapheme colour synaesthesia. The early prototype was based on a standard calculator, but it had a colour wheel attached. The person could assign the colour they perceived to belong to each of the numbers – creating their own personalised coloured calculator. After she used it, Berger's friend sent him an emotional message saying that it was the first time that addition and subtraction had made sense to her. Now maths was so much easier, and this had a huge impact on her life.

Berger further refined his device by setting up an experiment with 55 people with colour grapheme synaesthesia. Each person was allowed to choose their own colours on the calculator, and Berger measured their speed and accuracy in a mathematics test. Participants were very positive about the calculator and the majority of them were keen to use one in the future.

Berger acknowledges that this study was not blind, in that all the participants knew they were getting their own personalised coloured calculator, and it was based on self-reporting so experimenter bias must be taken into account. However, all the participants seemed to find it helpful and even if it had only a small benefit, Berger thinks it's worthwhile.

Anina Rich and her team have revealed other reaction differences in synaesthetes. In a study published in 2021 (the data was collected before COVID), the team took 18 grapheme colour synaesthetes into their magnetoencephalography (MEG) laboratory, where they measured the tiny magnetic signals produced by the subjects' brain activity. The patterns of brain activity showed that synaesthetic colour and colour we would all see overlapped – providing objective confirmation of synaesthetic colour experiences.

Interestingly, though, the synaesthetic colour showed up in brain patterns after a delay of about 100 milliseconds, compared with the neural signal when *non-synaesthetes* see colours. It demonstrates that what is happening in the brain when people experience synaesthesia is similar to what happens when we all see colour; it just happens a little later.

Hopes for further research

We know that certain regions of the brain specialise in particular functions, like colour perception, but no region works in isolation. Rich suggests that by understanding the bigger picture of how we perceive the world, we can start to help people who have a deficit or impairment in a specific area.

As an example, semantic **dementia*** patients lose their concept of an object. 'So they might look at a picture of a zebra and say, "Oh, it's a stripy horse." They can describe it but they've lost that concept. By understanding how we represent concepts of objects, how we bring together features and integrate information, we can potentially have a much greater understanding of clinical conditions like that. And I think synaesthesia offers us a unique insight into that system that we don't have through other means.'

(I'll have a lot more to say about dementia in Chapters 11 and 20.)

This isn't the only potential memory-related application. The evidence is still mixed, but it may also be that people could use their synaesthesia as a memory cue. They could, for example, use the colour they associate with a particular name to help them remember the names of people they meet. The name may be Debbie or Paula – but if the colour evoked is green, the colour the synaesthete usually associates with the name Debbie, then Debbie is probably the correct name.

Rich says there's also a lot more investigating to be done as to how synaesthesia might affect education. She's receiving queries from students and teachers wanting to know if there's a better way to utilise synaesthesia, to help their attention and learning.

'Synaesthesia is important to study,' observes Rich, 'because it gives us a unique insight into how perception might work and how we might represent objects. But I think more broadly it reminds us of the subjectivity of perception. So actually, you and I as non-synaesthetes kind of assume we see the world the same way. But we can never verify that …

'When we compare experiences with a synaesthete, we realise that we really don't see the same thing – but this is actually just an extreme of what is the case for everybody. We can never truly be sure that what we experience is the same as somebody else.'

The Taste of Purple: Synaesthesia and art

I'm listening to a jazz pianist and saxophonist playing music in the style of the great Duke Ellington and Johnny Hodges. Onstage with them, at her easel, is **Nina Norden** – known in the art world as Nina Nova – responding to the music as she creates a lively, colourful painting.

Nina is a synaesthete – and so was Duke Ellington. The renowned jazz pianist claimed he heard notes in different colours, while Nina says her synaesthetic perceptions of colour and shape are central to her creativity and art.

It's 2018, and I'm at a symposium called The Taste of Purple, organised by the Power Institute at the University of Sydney, with support from the Sydney University School of Medicine and the Synaesthesia Research group at Macquarie University.

After the performance I chat to Nina, and I notice even as I'm speaking to her that she seems to be constantly visualising and looking in different directions. She tells me she realises it can be annoying to some people that she's often not looking into their eyes as she's conversing but is paying attention to her range of synaesthetic perceptions.

Her art is very emotional, and she tells me she needs music to paint. The colours and shapes she chooses are her responses to the instruments she's hearing. Each instrument will have an aura around it that is influenced by the emotions of the musician playing it.

The music style significantly changes her art. 'If I'm listening to flamenco and Spanish music I paint a lot of red and very fiery, pointy shapes and like passion – things that I would connect with the music,' she says.

If she changes the style of the music she's listening to halfway through a painting, it can be confusing, because the colour and emotion in the painting change too much in response, so she knows she has to stick to the same kind of music until she completes the painting.

The connection between synaesthesia and creativity

Professor Anina Rich acknowledges that there is a stereotype that synaesthetes are involved in creative and artistic pursuits. A survey by her research group found that in a large sample of synaesthetes, 24 per cent were engaged in artistic occupations, ranging from visual arts, theatre and music through to architecture and graphic design, and that's compared with about 2 per cent of the Australian population generally. However, she points out that this could come down to sampling bias, and it's important to do more large studies before any firm conclusions can be drawn.

Mark Ledbury, the Director of the Power Institute, is passionate about the intersection of synaesthesia and art. As an art historian, he is fascinated by where creative impulses come from and what makes artists do what they do.

He says the artistic interest in synaesthesia really gathered momentum in the 19th and 20th centuries, when there were 'romantic attempts to look beyond what we might see as the limits of our perception. To tweak our perception so that we had a fuller understanding' of the world.

In the late 19th and early 20th centuries, synaesthesia was very much on the agenda of some of the great creators of the time, from French poets like Charles Baudelaire and Arthur Rimbaud to painters Wassily Kandinsky and Vincent van Gogh. Ledbury notes that during that period there was 'scientific interest in people who seemed to present anomalous gifts or aspects of their being that allowed them, for example, to see a letter or a number in a certain colour'.

He says there is little evidence about whether these artists were really synaesthetes, as there was no real means of testing this until the late 19th century. In the case of Kandinsky, argument still rages among neuroscientists, but the dominant view is that he probably *was* a synaesthete. 'He mixed, in his own writings and his own work, a strong sense of a kind of spiritual and cultural dimension,' Ledbury told me. 'So that's why he linked colours and tones and smells and sight.'

He adds: 'Maybe we wouldn't have certain types of music, certain types of jazz, certain types of painting, if it weren't for this kind of ability to enhance or alter perception. And that, to me, is fascinating, and an ongoing story.'

Mirror-touch synaesthesia: Joel's story

Joel Salinas is a doctor – which is a really interesting choice of career for someone who experiences the pain and emotions of other people as if they're happening within his own body!

Mirror-touch synaesthesia remains largely mysterious, and is of great interest to neuroscientists because it's believed to be an extreme form of a basic human trait. We all experience other people's pain and emotions to some extent, and this is referred to as empathy. It's known that watching someone being touched activates a similar neural circuit to actual touch. But for most of us, other parts of the brain intercept this circuit to tell us it's not actually *our* body being touched.

However, Joel's brain doesn't realise that what's happening to the other person is not happening to him, so he feels the physical sensation or emotion without being able to control it. As well as mirror-touch synaesthesia, Joel has grapheme colour synaesthesia.

About 1.6 per cent of the general population experience mirror-touch synaesthesia, and it's easy to imagine that it may be more debilitating than empowering. Further research into the condition may yield some valuable insights into the nature of empathy, the formation of self-identity and other critical aspects of social behaviour.

UK journalist **Helen Thomson** spoke to Joel for her book *Unthinkable: An Extraordinary Journey Through the World's Strangest Brains*. She told me about a conversation she had with him over dinner. He talked about injecting a needle into the spine of one of his patients, and feeling the same sensation in his own back.

Thomson asked Joel what he was experiencing right then. He said he was feeling the physical movements and emotions of all the people around him, like overheard conversations that you can tune in and out of.

Then he focused on Thomson herself. He said he could feel his face being scratched and she realised that she was absentmindedly scratching her own face. And then he could feel the brush of her hair, as if it were brushing against his own face, and the sensation of her lips crinkling on his lips. Thomson had a rush of awareness about how overwhelming it must be for people with Joel's condition!

Joel was a little hesitant to talk to Thomson about how his synaesthesia affects his relationships, because he was actually divorcing his husband at the time. He said it's very difficult when you're arguing with the other person and feeling *their* emotions as well as your own.

Relationships might be hard, but there are ways in which Salinas's mirror-touch synaesthesia may make him a better doctor. It allows him to have a high degree of empathy for his patients, so he's very well equipped to understand and help them.

And perhaps learning more about people like Joel can help us all become a little more empathetic. As Thomson puts it, 'Even these people with really incredibly extraordinary disorders or perceptions of the world, when we look at what's going on in their brain, we can see they just have an extreme version of skills and abilities that we all have, so it showed me what all of our brains are capable of if we could really understand them and utilise them in the best way possible.'

9

FACED WITH STRANGERS
Face blindness and super-recognition

Conversations with:

Dr Karl Kruszelnicki, Julius Sumner Miller Fellow, School of Physics, University of Sydney, science communicator and prosopagnosiac

Brad Duchaine, Professor of Psychological and Brain Sciences, Dartmouth College, New Hampshire

Emma Miall, prosopagnosiac

'That was Lynne Malcolm!': Dr Karl's story

I worked for many years in radio and science communication at the ABC with 'science guy extraordinaire' **Dr Karl Kruszelnicki**. We collaborated on science media projects and radio programs; we met and socialised at science events year after year. Yet for years, whenever we ran into each other in the office, or anywhere else, he didn't seem to acknowledge me. I felt quite awkward about it – not to mention perplexed by it – until finally I found out the reason.

Dr Karl has face blindness, more properly known as **prosopagnosia*** (ancient Greek for 'ignorance of the face').

'I grew up all through childhood thinking it was normal not to be able to recognise people's faces easily,' he told me. 'It might have been related to the fact that I was a refugee child from a non-Anglo-Saxon background growing up in a predominantly Anglo-Saxon suburb, and therefore I didn't have many friends and so there was nobody to recognise.

'But it came to me when I was a mature-age student at university studying medicine. And halfway through first year, two of my fellow students came to me and one of them said, "Hi Karl, what's my name?" And I said, "You're Stuart." And then the other guy said, "Well, OK, what's *my* name?" And I said, looking at him, getting a bit confused, "Well, you're Stuart too." And they said, "Don't you think it's a bit odd that we are *both* called Stuart?"

'And I looked at them and I could see that they were teenage and male and tall and slim, and they both had red hair. And I kept on looking at them and then suddenly it was as though an out-of-focus thing came into focus and I could see that one of them had glasses and the other one didn't, and I could see they had different chins. But I had to really struggle. And it was at that moment I suddenly realised I really can't pick the difference between one face and another, and presumably everybody else can, and there's something different about me.'

Karl told me that looking at faces is basically like looking at a bunch of bricks in a brick wall. He sees separate shapes of light and dark but gets lost because he doesn't see the face as a whole, as most people do. To compensate, he tries to focus on other characteristics, like how people walk and how they dress.

At the end of our conversation, I couldn't help but say: 'You and I have known each other through work over many years, and I do remember in the early years that you didn't seem to recognise me ...'

'No, I'm sorry, I'm really sorry,' he replied, 'but I wasn't being rude ... What you don't see is when I'm walking away I'm saying, "That was Lynne Malcolm! Lynne Malcolm, Lynne Malcolm!" And then I'll do it several times during the next hour and I'll write it down and I'll try and do it day after day.

'But then I meet somebody else as well and that sort of pushes you out of my memory banks. I've only got a finite memory. I wish I had a couple of terabytes of RAM in my head.'

After a number of years, though, he *did* train himself to recognise me, and now he *always* says hello!

How does face blindness happen?

For most of us, faces provide the prime way of connecting with the world and each other. Faces are the most important biological stimulus we see, and we've evolved to be exquisitely sensitive to the minutest changes in another person's face.

The average person has the ability to easily identify thousands of different faces, and pick out familiar ones in a crowd. For 1 in 50 of us, though, the brain's ability to recognise the human face breaks down. If the prosopagnosia is severe, it can be difficult even to recognise the members of your own family.

Prosopagnosia can be the result of some type of brain injury or neurodegenerative disease, or (as in Dr Karl's case) people can be born with the disorder and it runs in their family.

Brad Duchaine is Professor of Psychological and Brain Sciences at Dartmouth College, New Hampshire, and one of the world's leading researchers in face processing and visual recognition.

I asked him: how does it affect people's mental health?

'You know, it really seems to affect people differently,' he told me. 'Some people don't seem all that fazed by it, other people when I talk to them about the difficulties that they've had in life can really see that it has caused them a lot of problems and get quite emotional about it.

'So they see missed romantic opportunities, job opportunities. You can imagine you are interviewing for a job one day and you are meeting with somebody and then you blank them in the hall, well, that's not going to go over real well if that's the person making the decision about whether they are going to hire you or not.

'So it does cause increased **social anxiety** [a type of anxiety where people worry about social interaction], and it tends to cause **depression**.* As you can imagine, they tend to just be more withdrawn. They are less likely to be interested in going to a party, meeting people, because they have such a tough time in those situations.'

A stranger in the bed: Emma's story

When **Emma Miall** was 22, she was operated on because of a brain tumour. The right **temporal lobe** and right **hippocampus*** of her brain were cut out, which affected her memory, and as a result she now has great trouble recognising faces.

In one of the first facial recognition tests she did, she was shocked to find that she didn't recognise a photo of Princess Diana; she just saw very bland, generic features.

She gave me examples of when her face blindness has caused quite a bit of trouble and embarrassment. One time she woke up in bed with her boyfriend and suddenly she looked at him, asked him who he was and kicked him out of the bed. At that moment, in her head, there was a stranger in her bed.

Another time, when she was single, she was in a bar having a great conversation with a nice young guy. At one point he excused himself to go to the bathroom, and she waited for him to come back. A few minutes later a young man came over and sat down next to her with his drink, which she thought was very forward of him. There was a pause, and when she introduced herself to him, he was astounded and asked her what she was talking about; he said they'd just been chatting for ages.

Then Emma realised he was carrying his jumper and he'd taken it off while he was in the bathroom. She had been relying on his jumper to help her recognise him.

Clothing is one of the many cues she uses to compensate for her condition. 'I don't understand how other people do it,' she says, 'because so many people look the same. But I really have an appreciation for people who have an eyebrow ring, a cue, because I can remember piercings or tattoos because it's easy. It's like there's a template and I'll remember hair colour and things like that, which is why it's problematic when someone dyes their hair.'

Watching movies is quite challenging for her too, because she gets confused about who the characters are. Sometimes friends help her follow what's going on, but it becomes tedious, so she often gives up.

She muses: 'A friend of mine says, "Look on the bright side, Emma, you make a new friend every day." Which is great, but it's also kind of bittersweet, because it's harder to develop a close friendship … because when you treat someone with such unfamiliarity, even though you've seen them and spent quite a lot of time together, it doesn't send out positive cues.

'It's completely indiscriminate, you can't look at a face and have a great conversation and think, "I really like this person and so I'm going to study their face harder." It doesn't work that way.'

How we recognise faces

Brad Duchaine says the ability to recognise faces is quite complex.

'If I were to put you into a scanner and see how your brain responds to faces and objects, what I would find is there are a number of small specialised areas in the **occipital lobe** and the **temporal lobe** that respond much more strongly to faces than to objects.' Collectively these areas are known as the **fusiform face area** and are located in what's known as the fusiform gyrus. 'And we know based on work in monkeys that these areas are composed almost entirely of cells that respond to faces and faces only. And these little face areas, as they are called, are connected up to one another'.

Duchaine says that we process faces in stages. In the areas that are further back in our brain, we see a fairly simple representation of the face, like looking at somebody in three-quarters profile. 'And then as you move to more interior areas, so the areas that are further forward in the brain, you find that those areas are representing the face in what we call a view-independent manner', not tied to the particular image of the face you are seeing at that moment.

Recognition of faces is quite different from recognition of other objects. Duchaine suggests the reason 'might be simply that faces have been extremely important for millions and millions of years, and it's not just humans that have specialised machinery for faces, it's also other sorts of animals'.

Remembering faces

Some surprising recent research has shown that our ability to recognise faces actually improves as we get older.

This discovery was made by a team led by Professor Kalanit Grill-Spector at the Stanford University School of Medicine in California. They did brain scans of 47 people of different ages and, taking into account the individual differences in brain size, found that the adults had 12.6 per cent more solid brain matter in the fusiform face area than the children did. They suggest that these changes reflect how important facial recognition is for humans, particularly by the time we become adults, when we know thousands of people.

When I looked at **neuroplasticity*** in Chapter 1, I made the point that most changes in the brain throughout our lives involve **synaptic pruning**, a *reduction* in brain tissue. This exciting research shows that, as the number of people we recognise grows, the brain tissue in this region actually *increases*.

Coping strategies for prosopagnosiacs

In Chapter 1, I explained that in many instances where there is damage to one part of the brain, **neuroplasticity*** can compensate for that dysfunction. Sadly, this doesn't happen with prosopagnosia. It's a condition that lasts a lifetime. Therefore, people with face blindness need to develop coping strategies. Dr Karl and Emma Miall both work hard at compensating for their face blindness.

The late renowned neurologist Oliver Sacks revealed in his book *The Mind's Eye* that he had had difficulty recognising faces for as long as he could remember. He didn't think much about it as a child, but by the time he was a teenager it had become quite embarrassing for him and bewildering for his friends and peers. He learnt to recognise his two best friends by identifying very particular facial features.

Sometimes he even had trouble recognising *himself*. He described an incident when he apologised for bumping into a large, bearded man, only to realise that he was looking at a reflection of himself in the mirror!

His experience prompted him to delve further into this condition, and to write about the case study of Dr P, who lost the ability to recognise faces and other objects, in his popular book *The Man Who Mistook His Wife for a Hat*.

As he wrote in an article for *The New Yorker*, 'People with varying degrees of face blindness must rely on their own ingenuity, starting with educating others about their unusual, but not rare condition.' He noted that books, websites and support groups are cropping up to help people with prosopagnosia share their experiences and recognition strategies.

Brad Duchaine suggests that one way to improve face recognition in those with face blindness is to train them on recognising a small set of faces that are really important to them, and give them a *lot* of practice. He gives the example of Bill Choisser from San Francisco, whom he's met during his research. Choisser said that by Bill Clinton's second term as President of the United States he started to be fairly good at recognising him. It came with heavy practice!

While many adults with face blindness develop compensatory strategies, these techniques can be too sophisticated for children. This is prompting researchers around the world to develop intensive training programs that they hope will significantly assist these children's facial recognition skills.

The other end of the scale: Super-recognisers

At the other end of the spectrum are people with what is called **superior face recognition ability**. It's genetic, so the skills of these 'super-recognisers' are encoded in their DNA. Interestingly, research so far shows that this incredible ability is quite independent of intelligence, personality and other cognitive and perceptual skills.

Duchaine explains that super-recognisers 'might see somebody just in a brief encounter and then run into them several years later somewhere else and they are able to recognise that person. And so they are able to do things that certainly most of us can't even imagine. And they find they have to actually restrain themselves when it comes to revealing that they recognise others because people find it kind of creepy when somebody says, "Oh yes, I saw you walking across the quad three years ago." It sounds like a stalker.'

While studying prosopagnosia more than a decade ago, Duchaine started to hear from people claiming to have the opposite condition, extraordinary facial recognition skills, so he and his colleagues Richard Russell and Ken Nakayama decided to put these people's skills to the test. They published a paper in 2009 about four of these people who were off the charts, and since then they've heard from a lot more super-recognisers with amazing abilities.

'One of the things that a lot of them reported when they watched movies and television shows, they are frequently recognising the extras. So they say, "Oh, I saw that person in this other show." And of course you and I would be totally oblivious to the fact this person was someone we had seen before.'

One of the most interesting things, Duchaine says, is that the London Metropolitan Police are putting these super-recognisers' skills to use. In 2015 they formed a super-recogniser unit – a team of over 200 people with exceptional facial recognition abilities who match up criminals with mugshots in the police database, and identify faces that are captured on CCTV. 'And they've been able to identify more people from images now that they are using these super-recognisers than they are from either fingerprints or DNA. So it's been really effective for the Met, and hopefully it's going to be something police departments around the world are going to be starting to make use of.'

Duchaine thinks face processing is a good example of how natural selection has endowed us with a highly specialised brain mechanism to carry out a very specific and important evolutionary task.

10

SEEING WHEN YOU'RE BLIND

Charles Bonnet syndrome

Conversations with:

Anne-Gabrielle Thompson, person with Charles Bonnet syndrome

Ross Brown, person with Charles Bonnet syndrome

Mary Lovett, person with Charles Bonnet syndrome

Professor Chris Plummer, Neuroimaging Department, Swinburne University of Technology; Consultant Neurologist, St Vincent's Hospital, Melbourne

'Like a video unravelling'

Anne-Gabrielle Thompson, 91, is blind. One day a few years ago she was being driven back from a Toastmasters meeting by her friend – and the most unexpected thing happened.

She looked out the window and saw piles of timber stacked up along the side of the road. How could this happen? She was blind!

A few weeks later, she was talking to her daughter at the dinner table, 'and suddenly this face appeared in front of me. And whenever I moved my eyes, the face was there. I couldn't say whether it was a man or a woman; it was a very handsome face with a very high forehead, longish wavy grey hair. And when we finished eating, the face disappeared.'

The third time it happened, Anne-Gabrielle saw lots of little children in pink and white striped pyjamas, waving and wriggling their bodies around. She went on to have more of these strange hallucinations, 'like lots of little insects crawling on my dining table'.

* * *

Ross Brown, 87, recalls: 'Shortly after I became blind I looked out the window one morning to the house next door and the house was not there, there was a huge pile of rubble with a lad about eight or nine years old running around on it.'

Another time, he was sitting at the breakfast table and became aware of a well-dressed man sitting next to him. 'But there was no one there, it was an image, possibly an image of my brother, who died quite a few years ago.'

* * *

Mary Lovett, 87, is also blind, and described to me a vivid scene she would see regularly.

'It was like a video unravelling. Everybody was beautiful. The men were wonderfully dressed and really good-looking, and the girls were gorgeous – wonderful hats, wonderful hair, beautiful dresses and wonderful complexions.'

She saw this scene over and over again for quite a few years, and found it fun – until the day when one of the people winked at her. Then she thought, 'This is not a fun thing, this is something else!'

* * *

Anne-Gabrielle, Ross and Mary all have what's known as **Charles Bonnet syndrome (CBS)***. ('Bonnet' is pronounced 'Bon-ay'.) Sufferers have some form of acquired vision loss, and are otherwise sound of mind, but experience phantom images and visual hallucinations.

The syndrome was first described in Geneva by naturalist Charles Bonnet, in a paper he wrote in 1760 about his grandfather, who had impaired sight and unusual visions. It took nearly 200 years before the condition was named after Bonnet by Geneva neurologist George de Morsier.

It's estimated that 25 per cent of patients with significant visual impairment have had at least one CBS episode. That equates to around 125,000 Australians. But the actual number of sufferers is almost certainly higher, because the condition is often misunderstood and misdiagnosed.

The late neurologist Dr Oliver Sacks took a great deal of interest in this syndrome, suggesting that more research into the phenomenon would give us valuable insight into how the brain works. Sacks also experienced the syndrome himself. After he had become visually impaired – blind in one eye, with impaired sight in the other – he would see geometrical hallucinations and often draw pictures of them in his notebooks. He even wondered whether cave art and ornamental art might have derived from these experiences.

What causes it?

Professor Chris Plummer, a head of neuroimaging research at Swinburne University and a neurologist at St Vincent's Hospital in Melbourne, explains: 'It's a condition that occurs with visual impairment from a lesion or abnormality anywhere along the visual pathway, so it can be from the eye, as in macular degeneration, to the **visual cortex** [at the back of the brain], as in a stroke.'

For a diagnosis of CBS, three criteria must be met. Firstly, the hallucinations must be recognised by the person as not real. Although initially they may be confronting, they are phantom visions – as opposed to delusions, when the person can't distinguish the illusion from reality.

Secondly, the hallucinations must be purely visual, like a silent movie only in colour. Other sensory phenomena are absent. 'And this really adds to the oddity,' says Plummer, 'as patients can be confronted by typically noisy scenes, like some examples of speeding trains across the living room wall'.

The third criterion is that the hallucinations cannot be linked to any brain impairment such as Parkinson's or **Alzheimer's disease*** – although some sufferers do initially feel they have **dementia,*** because the hallucinations are so unusual.

Neuroscience cannot yet explain exactly how or why CBS images occur. However, Plummer says that some compare it with phantom limb syndrome – where a person can still feel a limb that's been amputated – but without the pain.

The hypothesis most scientists favour is **deafferentation**, the loss of electrical signal to a group of nerve cells. With vision loss, particularly if it's sudden, there will be a decrease in stimulation to the receiving cells along the visual pathway, which can cause a spontaneous burst of electrical activity and give rise to the visual hallucinations. 'But it's still not fully understood,' Plummer explains.

It's also thought that the nature of the hallucinations is dependent on the specific area of the visual pathway that's affected. If it's the **primary visual cortex**, the hallucinations are more likely to be simple, involving lines, dots, dashes and geometric patterns. When the **secondary visual cortex** – which mediates colour, facial recognition and intricate movements – is affected, the hallucinations are likely to be more complex.

Plummer told me: 'Of the patients I've seen, probably the most common themes I've come across would be of insects, often giant insects such as spiders and ants crawling on walls and tables, like Anne-Gabrielle's description. Children are common, usually running around in brightly coloured clothes or outfits like Anne-Gabrielle and Ross described. Objects with very prominent geometric shapes, like brick buildings, trees, huge timber planks, like Anne-Gabrielle and Ross described.

'Landscapes are also not uncommon. Usually they are very scenic and inviting. Faces are one of the more common visual hallucinations that are seen, often with very prominent eyes and teeth, and those areas in the brain are well represented in the secondary visual cortex. Anne-Gabrielle and Mary had some of these experiences.

'The other two visual episodes that patients can experience are miniatures, so they see little soldiers or Romans marching or riding in chariots, often with prominent headdresses, things like elves and fairies on tiny trains or cars. And then you come down to the, you might say, downright bizarre, with things like goats walking through the lounge room wearing overcoats. I had one patient who described the carpet rolling up in front of them to form this giant anaconda snake and then it just slithered out of the room.'

Although some may find these Charles Bonnet hallucinations confronting at first, once they understand what they are, they're not usually disturbed by them – in fact, they can be fascinated and entertained by them.

However, Mary Lovett told me that when she began to be troubled by the hallucinations, she joined a support group through the Charles Bonnet Syndrome Foundation in Melbourne.

Many people with the syndrome are worried that others will think they're crazy, so they keep it a secret. Mary hadn't even told her husband, so sharing her experience in the group helped her to accept it and open up to family and friends.

How is it treated?

There is no cure for CBS, and medication has not been found to be reliable. 'The problem is that a lot of these medications have side-effects that can be worse than the actual syndrome,' says Plummer.

He suggests early diagnosis is helpful. It's also important to explain to the patient that CBS isn't linked with psychiatric illness, and usually recedes over time.

'There are factors, lifestyle factors that can diminish the intensity of the episodes,' Plummer says, 'such as avoiding sensory deprivation really. So the classic scenario is the elderly patient who is living alone in a low-lit house with a fairly sedentary lifestyle. So anything they can do to improve environmental stimulation, enhance environmental stimulation, can diminish the frequency of these attacks. Strengthening social networks can help. Engaging in distracting activities like more regular physical exercise can help.'

Some investigators have found that deliberately moving the eyes around and scanning the room during the hallucinations can help, and sometimes closing the eyes can shorten the episodes.

As well as this, Plummer says, 'anything that can be done to optimise the patient's vision can help. So there are reports of patients who have had cataracts and experience Charles Bonnet syndrome and had surgery, and their Charles Bonnet hallucinations have vanished with surgery.'

Plummer sees great value in researching Charles Bonnet syndrome further: 'The point is that we don't fully understand how the visual system works in its entirety in health. We don't really understand how the brain sees, as the networks involved are enormously complex; we certainly don't fully understand it. And as with many conditions in neurology, key insights into how these systems work can be gained from situations where something goes awry somewhere in the network, as in Charles Bonnet syndrome.'

Part III

NEW INSIGHTS INTO
DEBILITATING
MENTAL CONDITIONS

WARNING: This part of the book includes content that some readers may find disturbing.

11

DEMENTIA

Conversations with:
Dr Jules Montague, consultant neurologist
Lisa Genova, neuroscientist and author
Dr Kailas Roberts, Director, Your Brain in Mind, Brisbane
Christine Bryden, author and early-onset dementia sufferer

My family's experience of dementia

How sadly ironic it was when it became evident that my father –
a kind, patient, sharp-minded, logical man – had developed
vascular **dementia*** (associated with a lack of blood circulation
in the brain). Suddenly he was losing track of conversations,
getting lost, needing help from my mother to dress and do so
many other things …

He was often irritated and highly frustrated with himself, but
he had also become softer and more emotional. His eyes would
fill with tears at the drop of a hat, and he began to long for a pet
dog. Not just any dog, he wanted a Great Dane! Well, that didn't
quite happen – but jet-black Louie was disproportionately large
for a Labrador pup, which meant that Dad had just about got his
wish! She was a big, spoilt, adorable sook.

Dad's dementia worsened, and it became harder and harder
for my mother to look after his every need while taking care of

herself. It was becoming obvious that he would have to be put into a care facility.

Then one day he had a heart attack as he was accompanying Mum to the letterbox. He was taken to the hospital, and with Mum and my brother and his family by his side he talked agitatedly about needing to catch the fast train to his hometown, the place of his birth. By the end of the day a final heart attack had killed him.

My partner and I were out of the country when I got the news that my beloved dad was dead. I lay awake all night in disbelief, sadness and emptiness.

When we arrived back in Australia my brother and I went to the funeral home to say goodbye.

Having cared so intensively for Dad, Mum was left depleted and shattered after he passed away. She'd forgotten who she was and lost any sense of her own purpose in life. After a while she managed to pull herself out of the hole to some extent, but it wasn't long before she began to close down both physically and mentally, to the extent where she couldn't be looked after at home. Our family arranged residential care for her, which she passively accepted.

She deteriorated quickly after she moved into the home. She grew more and more uninterested in participating in social activities and became extremely weak, to the point where she couldn't get out of bed on her own. After a short period of struggling to get her words out, she lost the power of speech. She too had developed a type of dementia.

Mum had been a kind, loving, funny and mischievous woman. Visiting her after she stopped speaking was excruciatingly sad. She seemed perplexed as she stared deep into my eyes, then her gaze would shift and escape right through and beyond me, to some other dimension I would never be privy to. Sometimes I detected a glimmer of a tear filling her eyes in response to something I said, or a half-hearted raise of the eyebrow: a cheeky gesture she used to often use.

Out of desperation, visiting one day with my brother, I tried another way of reaching her. I had learnt in my research about the powerful role music can play in tapping into memories and emotions from the past. (I'll tell you more about what's being done in this area in Chapter 22.)

Might playing music from her past work in reaching my mother?

We decided to play her one particular song we knew had been one of Mum and Dad's favourites. The classic 1950s track 'Fever', sung sensuously by Peggy Lee. They would have been listening to this music in their early 20s.

I streamed the song from my phone, which I placed on the arm of her recliner. My brother and I did our best to suppress our grooving and foot-tapping – but I would be lying if I said we detected any sign of recognition from Mum.

What was she feeling? What was she thinking? Did it mean anything to her that we were there? It certainly meant a lot to us.

I feel deeply privileged to have been at her bedside during her final breaths. My daughter and I sat and stroked her hair, sharing precious memories, singing songs she loved and reciting poetry she used to read to us. We feel sure that she knew we were with her until the end.

Each experience of dementia is different – but every one of those experiences is devastating for the person affected, for their loved ones and for society at large.

What is dementia?

UK neurologist **Dr Jules Montague** vividly recalls diagnosing her patient Anita with a form of dementia. Anita asked her: 'Now that I have dementia, am I still going to be the same person?'

'And these are questions of identity, sameness, the real person,' observes Montague. 'And we only tend to, I think, ask them when something goes wrong with us.'

My family's struggle with dementia is far from unusual. It's estimated that around 450,000 Australians are living with some form of the condition, with over 1.6 million people involved in their care. But many of us have a limited understanding of exactly what dementia is.

The first important point to make is that dementia is not one specific disease, but an umbrella term for over 100 different neurological conditions. It is characterised by a persistent decline in intellectual abilities such as memory, thinking, judgment, language, behaviour and everyday life skills. **Alzheimer's disease***
is the most common form, affecting 70 per cent of those with dementia. The causes of Alzheimer's are still not fully understood, but the end point involves a buildup of **beta-amyloid plaques** (deposits of the **amyloid protein**) and **neurofibrillary tangles** (formed by abnormal **tau proteins**).

The second thing that's important to note is that while our memory declines in our 50s and 60s (as explained in Chapter 3), dementia is *not* a normal part of ageing. However, it does mostly occur in older people, with around 10 per cent of people over 65 experiencing it. Dementia was quite rare before the 20th century, because people didn't live as long, so the disease didn't have time to develop.

One of the most devastating things about the disease, as I saw with my parents, is that it really does eat away at people's identity. Our memories form the basis of our sense of self. So if our memory is damaged, what remains of the person left behind? Does it mean we are no longer ourselves?

These are some of the fascinating questions UK neurologist Dr Jules Montague has explored through the moving experiences of her patients. These experiences inspired her to write her book *Lost and Found: Memory, Identity and Who We Become When We're No Longer Ourselves.*

She's concluded that sometimes, in the depth of the loss associated with dementia, there is profound hope and promise. After Montague's patient Anita worried about whether she would be the same person with Alzheimer's dementia, 'she fought to remain as much herself as she could, and her family fought as much as they could to see who she was … We have this perception of Alzheimer's and other forms of dementia that people lose themselves completely, and Anita, for most of her journey at least, proved that that wasn't the case.'

In the end, Anita lost the ability to speak. As I know from my experience with my mother, it's really challenging, after language goes, for loved ones to think about the person in the same way.

Often as some forms of dementia progress people's grammar and syntax get mixed up, and it becomes difficult to follow their stories in the chaos, 'but perhaps that chaos and repetition is our problem and not the person with Alzheimer's,' Montague says. 'What I'm saying is perhaps not only are we more than our memories, but we are more than coherent sentences.' There are lots of other ways of telling stories: we constantly communicate through our habits, gestures and smiles.

In **semantic dementia**, people lose their language ability. Montague describes one semantic dementia patient of hers who became completely mute. His wife was at a loss as to how to help him, but one day she gave him some pens and pencils, and he started to draw. He required 24-hour care for everything else, but at the same time every day he would display a beautiful creative talent his wife had never seen before. Apparently, even though the front of his brain, which controls language, began to shut down, the visual perceptual, artistic part of his brain became free to roam. 'So it's something incredible, when you think that all has been lost, there can be some discovery there and some incredible artistic talent.

'I think one of the problems in medicine as a doctor is we ask what people *cannot* do,' Montague adds. 'And I think therefore we lose something very valuable when we speak to patients when we only focus on the *can't* and the *cannot* rather than the *can*.'

She also says that there's a tendency, if an Alzheimer's patient is disoriented, to bring them back to the here and now by asking them what date it is, what year it is etc – but there's an understanding now that perhaps it's better to help the person go back to whatever state feels safest for them.

A change of personality: Martin's story

Martin was very compassionate, well liked in his small village in the west of Ireland, and involved in various committees. He was very organised, very polite.

But over the course of a couple of years he became a completely changed man. He would stagger down the street, shouting and gesticulating.

Montague was his doctor, and diagnosed him with a type of dementia that affects the front of the brain called **frontotemporal dementia (FTD)**. The FTD had completely changed his personality.

'And that led me to think about why it is that personality change threatens our identity so much,' Montague told me. She found some 2015 research from Yale University that collected data from family members of patients suffering from neurodegenerative disease. It showed that it was changes in *moral behaviour*, not memory loss, that caused loved ones to say that the patient wasn't the same person any more.

'So if you can think of someone who is very honest, who has a lot of empathy, a lot of compassion in your life, and now you can imagine they wake up the next morning and they don't have any of those traits, you will probably think they are a different person. If, however, their *non-moral traits* change, so if they were less creative or less curious, you wouldn't necessarily feel that they had changed intrinsically. And Martin's story really taught me because he lost those things that were so integral to who he was, the empathy and so on, and he became a different person. And that's what people often report in this condition called frontotemporal dementia, a changed person, a different identity.'

Martin's niece Tara told Montague that in the past he had always been empathetic towards her and her mother, but since the dementia he no longer seemed to care, even though her mother had been diagnosed with cancer. 'Martin just didn't seem to register it.' The family began to wonder whether *this* was the real Martin and the compassion he had shown in the past had just been a veneer.

'And towards the end it took me a long time to try and figure that out, and I really believe that people shouldn't be defined by the destruction of their brain. So it feels unfair to me that we would say this is finally the real Martin who has now been exposed by this brain condition. And I would like to say that we should preserve Martin as we remember him best rather than judging him at his worst.

'We often think about identity as being binary – it's either there or it's not there, sameness is either present or absent – but that's not the case at all,' Montague observes. 'I still think it's important to look for those semblances of sameness, particularly if there's someone in your life – and I know this is difficult, but if there's someone in your life who has some sort of brain condition ...

'I think looking for those semblances of sameness along that path is worthwhile, because sometimes you can make some wonderful discoveries and find remnants of the person that you once loved and that you once knew. And I think it's an important journey to take.'

Still Alice: Lisa's story

American neuroscientist-turned-author **Lisa Genova** became intrigued by dementia as she watched her grandmother live with Alzheimer's disease – so much so that it inspired her to self-publish a novel, *Still Alice*. It's the story of a 50-year-old professor of linguistics who develops early-onset dementia. It became a bestseller, and Hollywood actor Julianne Moore won an Oscar in 2015 for her moving portrayal of Alice Howland in the film adaptation.

Lisa Genova told me that she got up to speed on the molecular biology of Alzheimer's and read many self-help books, but none of that helped her, as a granddaughter. The thing that remained elusive was the same thing I was desperate to know about my mother: what does it actually *feel like* to have this condition?

'And as I spent time with my grandmother I was desperate for an answer to that because that was what I needed to stay connected to her.

'I was so heartbroken by what was happening to her, and I was very unnerved and baffled by the various odd behaviours. She was talking to a stuffed animal named Henry, she was trying to breastfeed these plastic baby dolls, she was getting lost in her house, couldn't find her way to the bathroom, having accidents, looking at the numbers at the front door of the house and saying, "Oh, my gosh, this is where I live." I didn't know how to be with her with her Alzheimer's.

'And so this idea of what does it feel like, I made this intuitive leap that fiction was the place to answer that, so I thought, some day I'll write a novel about a woman with Alzheimer's and tell it from her perspective.'

Genova's central character, Alice, has early-onset Alzheimer's. This type of disease is strongly linked to your genes. If you inherit one of the three gene mutations associated with early-onset dementia, you will probably have symptoms of the disease. Fortunately, though, these devastating cases make up less than 5 per cent of all diagnoses.

It's not clinically any different from later-onset Alzheimer's, except that the symptoms begin before the age of 65. What *is* different, though, is the impact such a condition has on the life of a much younger person.

'So you can imagine most 85-year-olds are probably not working any more, so they are not accountable to a corporation or a boss. And so you can make more mistakes as an 85-year-old and it's less alarming to you and the people around you.

'If you are 50 years old and you have a big job, and maybe you have teenagers and you have lots of responsibilities and are accountable both professionally and personally at a higher level, the slightest mistakes that you make in cognition, language and memory are going to be exquisitely noticeable to you and the people around you.'

Genova reflects: 'I think that so many of us place our worth and identity in what we do ... And for Alice, she is a Harvard professor, so she has placed all of her worth and identity in her ability to think and remember and her language skills. And so now she has a disease that is going to rob her of those very things that she has placed all of her identity in. And so one of the ways in which she actually grows as the Alzheimer's is diminishing her is [in] understanding that she is more than what she can remember. That even if she can't remember her basic biographical information, like her address or her daughter's birthday, she can still love her daughter, she can still matter to her family.'

In Genova's correspondence with people about Alzheimer's, she noticed the word 'still' in every single email. 'I'm still here', 'I still love my husband', 'I still love to garden', 'I still want to contribute', 'I still want to see my children grow up'. Hence the title of the book, *Still Alice*.

As part of her research, Genova connected with 27 people with early-onset Alzheimer's, and even stayed in touch with some of them. It's been an amazing experience for her. 'They taught me what you can't learn in the textbooks and what you cannot learn from the neurologists, and they really shared with me their most vulnerable selves. It was such an honour to witness and understand what people with Alzheimer's go through.'

She's really proud of how her book and the film adaptation have turned the spotlight onto this condition: 'Truly Alzheimer's has been in the closet for a long time, and I get that we didn't want to talk about this because it was scary and there's so much shame and stigma attached to this disease. So people going through this tend to go through it in isolation. They are excluded from the community, and we haven't talked about it. It's awfully hard to cure something that seemingly doesn't exist ...

'*Still Alice* has been part of the progress that has led to conversation. So both the book and then the film in a very big way have dragged this topic out of the closet and into people's living rooms … And so we are hearing this global conversation about Alzheimer's and we are seeing an increase in funding. And so I know that we are on the right track.'

Treatment, delay and prevention

As well as the massive impact on sufferers and their loved ones, the public health burden of dementia is considerable. The worldwide prevalence is projected to nearly triple by 2050, because of ageing and population growth. In Australia it's the second biggest cause of death, and cases are projected to increase to more than 1.1 million before 2060.

Having dementia in my family, and knowing that there is a genetic component to the condition, naturally I wonder whether I or my children will be affected by it. And I know I'm far from alone.

So, the race is on to learn more. Researchers around the world are heading in a range of promising directions to find the causes, effective treatments and ways to prevent the condition. Many others are focused on gaining better insights into the experiences of those with dementia, in order to improve their care and optimise their quality of life.

Dr Kailas Roberts is passionate about raising dementia awareness. He's a psychiatrist and dementia expert and director of the Your Brain in Mind clinic in Brisbane, where he's focusing on the best ways to treat and prevent dementia, as well as delay its development. He is also the author of *Mind Your Brain: The Essential Australian Guide to Dementia*. I spoke to him recently, following the release of his book.

Maintaining a healthy brain

Roberts told me that over the last decade or so scientists have learnt more about how to prevent or delay the onset of dementia through various lifestyle factors. He cites a 2020 report commissioned by *The Lancet* medical journal that concluded these factors could have an impact in up to 40 per cent of dementia cases. It's unlikely that one single thing is going to make a difference; a multi-pronged approach is required. But the earlier you start, the better. Mid-life is a good time to begin thinking about the changes you can make for better brain health.

The pillars of good physical and mental health, such as a healthy diet, regular physical exercise and high-quality sleep, are a great place to start. Avoiding smoking and excessive use of alcohol over the long term is better for the brain too. **Dr Norman Doidge**, whom I introduced in Chapter 1, told me about a study by the Cochrane Institute in Wales that showed that five activities – exercise, not smoking, not drinking more than a glass of wine a day, eating four servings of fruit and vegetables daily and maintaining a normal weight – can reduce the risk of developing dementia by a staggering 60 per cent.

Social connection is also very important. (I'll talk a lot more about that in Chapter 23.) Roberts told me about a 2020 report by the National Academies of Sciences, Engineering, and Medicine in the United States that found nearly a quarter of adults aged 65 and older are socially isolated and lonely, which puts them at greater risk of dementia and many other health problems. Mental illnesses such as **depression*** and **anxiety,*** particularly if they're severe, chronic and left untreated, can also increase the risk of dementia.

Roberts says that the *Lancet* study found the biggest modifiable risk factor for dementia at a population level is poor hearing. So it's critical to check your hearing on a regular basis.

Maintaining complex mental activity is also helpful. Roberts describes this as just using your brain in novel ways – putting the brain under a little bit of pressure, but not to the point of being distressing. Doing crosswords, for example, can be helpful, but if you love crosswords and you do them all the time they won't really challenge your brain. It's better to try something that's new and different for you, to promote neuroplastic changes and increase what's called your **cognitive reserve**. (The tips for nurturing your creativity on pages 53–54 could be helpful.)

It's extra-beneficial if you do these activities with other people, so you are getting social interaction. It's also helpful if they are activities involving movement and motor function, such as dancing or tai chi.

Roberts points out that there is a lot of controversy around phone apps that claim to help you improve your cognition. He says the jury is still out on the benefits of these, though some research findings have been promising. One of the issues is that you may improve in the game or task you do on the app, but it's uncertain whether that translates into meaningful brain change. The most effective approach is just to do something a bit different, as I outlined above.

Meanwhile, Roberts has developed his own phone app, BrainScan, to help people identify their dementia risk factors and find ways to address them.

Reasons for optimism

Roberts is optimistic about what the future holds for people with dementia. He told me that while there's no cure yet, there's been some excitement in the last year about a drug that clears the **amyloid plaques** associated with Alzheimer's disease from the brain. While there's some controversy around whether this drug will improve the clinical symptoms of the disease, Roberts says it's an encouraging time in the field, with other new drugs also being investigated that could be helpful, especially if they are used earlier, before dementia has fully developed.

The idea of a multidisciplinary approach is also gaining traction. In his clinic, Roberts works with occupational therapists, physiotherapists, dietitians and psychologists to offer dementia patients a broad range of support.

I'll mention some of the other exciting work that's going on to enrich the lives of dementia patients in Part IV (particularly Chapter 20).

Before I Forget: Christine's story

When **Christine Bryden** was diagnosed with early-onset dementia at just 46, she decided that she needed to tell the world about her inner experience before she forgot it. That inspired the title of her book *Before I Forget*, one of several she has written since her diagnosis.

She told me how she felt when the doctor first told her. 'I am too young, it can't possibly be this. And he said, "No, no, you've got about five years until you are demented, and then another three until you die." And, I mean, it was just unbelievably cruel.'

She observes: 'To me, the toxic lie of dementia is what you are told at the beginning … It's not only just a lie but it is toxic because it becomes a self-fulfilling prophecy, and you retreat into **depression*** and despair, you fulfil that prophecy …

'We need to just live every day as if it's our best day, even if it is our last day, and make the most of every moment in the day and make it a moment full of wellbeing and not think about the horrible future that might await us, because that *does* become toxic.'

That conversation with her doctor occurred over 20 years ago, and it inspired her to devote her life to advocating for better understanding and destigmatisation of dementia. In 2003 she became the first person with dementia to be elected to the board of Alzheimer's Disease International.

'I've come a long way in that 20 years. At the beginning it was dreadful, it was just so miserable, there was no support, and I couldn't rise above and think positively. But reflecting back has been such a different experience.'

Christine was a very intelligent and highly motivated young person, proud of her sharp memory. But by the time she was diagnosed, she had already experienced several huge challenges in her life. As a teenager she developed anorexia nervosa, but worked through that to become a biochemist. The early warning signs were there, but she ignored them: 'Looking back, I recall in my late 30s getting confused and lost in the middle of Sydney on a familiar drive to CSIRO laboratories … And then getting names a little bit muddled. But it was abnormal for me. I knew something was wrong, but I kept on putting it down to stress.'

Eventually she took on an even more stressful role as advisor to the Prime Minister. By that time she'd married and had three lovely daughters, but very sadly it turned into an abusive and violent relationship.

'And I was just getting all these thundering migraines, and I thought it was just because of work life and home life and stresses and trying to keep everything going, so I was just ignoring them and taking tablets, as you do.'

Eventually she found the strength to leave her husband. Without giving him warning, she packed up a removals truck and escaped with the girls, the cat and the budgies. 'And during that year afterwards I was still finding all these so-called signs of stress and the migraines, and that's when I finally went to the GP, and she finally sent me for brain scans.'

When she received the diagnosis, 'my whole planet tilted on its axis. It felt like an earthquake under my feet. It was just horrific. I couldn't think of the future without horrible fear.'

When Christine and I spoke in the ABC studios she felt lucky that what she calls her 'brain disappearing trick' is not progressing as fast as was expected: about 4 or 5 per cent a year. 'That is giving me time to try and counteract it. For other people too, they might not have been operating at such a high level beforehand, or might not have had such a lot of what's called **cognitive reserve**,* you know, studies and games when they were a kid … I mean, we used to play IQ tests for fun, my mum and I.' Christine really is the embodiment of Dr Kailas Roberts's sound advice in the previous section.

However, her condition is 'definitely progressing'. She told me: 'Even walking in, I was trying to remember who on earth I was meant to be talking to, so I had to write it down. And whether I'll remember tonight, I don't know. Possibly not. It's very peculiar; let me say it is very peculiar.' She wouldn't be able to do all that she does without her second husband and carer, Paul Bryden.

She told me how they got together: 'So it was nearly two years after. And I just felt, look, I'm just going to have to get out there and meet people and just start socialising and be more positive. Then I did the even more ridiculous thing, of course, because I was lonely, I went and joined an introductions agency. That really made me feel more positive, because then I met Paul.'

She laughed as she told me that she announced within an hour of their first date that she had early-onset dementia: 'And he just said, "Oh yes, my father had Alzheimer's, he died a few years ago." He just carried on as if I hadn't said something totally earth-shattering. And he's still here, still supporting me, still being my enabler, helping me do as much as possible, like getting me here to speak to you, and helping me travel and telling me what day it is and what it's going to be tomorrow and what I did yesterday. He just keeps me oriented so I can keep going. He is helping me right the way through to do as much as I can, *while* I can.'

Together, they've learnt a lot about caring for someone with dementia. 'First of all let me say that the people who care for people with dementia are wonderful … But really I think every single moment with the person with dementia should be seen as a time of being able to connect. So, for example, instead of just dragging me into the shower, just talking with me and connecting with me, looking at me in the eye and just chatting about the shower and the day, just gently talking me through it …

'I think we need to change that whole attitude, that I'm going to be just a physical object of care. I'm always going to be fully human and fully present, I'm not going to be just Mrs So-and-so sitting over there in the corner who needs a shower, I'm going to be someone with a life story, someone who can be connected with, even if I've got no words.'

Bryden describes her inner experience now as a journey of being present: 'It's hard to hold on to the future. And I've likened it in one of my talks I think as being like on a carpet that's all rolled up in front of you and is rolling up behind you and you're just standing on the carpet right there.'

She explains, 'we somehow discard people with dementia as just people with dementia, and yet they have discovered that treasure of living in the present moment. It's how we have to be. We don't remember yesterday or know much about tomorrow, but we are present.'

While listening to Christine, I found myself hoping that this was my mother's inner experience too.

12

ANXIETY

Conversations with:
Willow Freer, who has experienced anxiety
Miriam Heatherich, Willow's mum
Dr Jodie Lowinger, clinical psychologist and CEO of
The Anxiety Clinic, Sydney

The 'Worry Bully': Willow's story

When I met **Willow Freer** and her mum **Miriam Heatherich**, Willow was 10 years old. She'd been suffering from **anxiety*** since the age of seven and a half, when her family moved to Australia from Singapore and she started at a new school.

'I was scared of making friendships with new people, I was scared that I was going to embarrass myself. I was worried that people would make a lot of fun of me and I'd never be able to get friends at that school and I would get bad results because I was so focused on friendships.' Some of the other girls teased her. 'That made me feel really sad, and so I'd usually sit by myself.'

She described what the anxiety was like. 'I felt really tense, and my palms, they would sweat a lot when I was writing stuff and thinking about it, when I was doing work in the classroom and stuff. And my brain was all foggy.'

Eventually the teacher put a stop to the bullying, but it didn't put a stop to Willow's worries. 'I'd be worried about my grades and if I was doing OK or not because I was getting a lot of Cs and just Cs, no Bs, no As then, and so I got really worried about that. I also got really worried about how I slept because I didn't sleep very well then, and I got really worried that maybe if I didn't sleep very well, I wouldn't do very well at school.'

'In retrospect even as a baby she was highly anxious,' says Willow's mum, Heather, 'and anxiety is very prevalent on my side of the family. So if you have an anxious mum and an anxious child, it's definitely a recipe for extreme stress.

'So when we moved back to Australia after five years in Singapore, I underestimated how impactful those changes would be. New house, new school, new friends, new weather. And I was in my own world of stress and pain, and I just noticed that Willow was not going to sleep, and at that point at seven and a half years old she was up until 10 and 11 some nights, although we would get into bed at 7.30.

'I felt completely helpless actually, I had access to no tools to help a seven-and-a-half-year-old, and that, as a mum, it's so challenging. I'm sure I can appeal to all mums: if you can't help your child you feel useless yourself, so then it just compounds the problem.'

That was when they turned to clinical psychologist **Dr Jodie Lowinger** at The Anxiety Clinic. Lowinger has quite a number of clients around Willow's age. It's estimated that 25 per cent of the population develop clinical anxiety, and more and more of them are children or adolescents. Lowinger explains: 'Kids are experiencing anxiety that is, certainly in Western society, largely around some of this digital technology that is overwhelming now, with everybody presenting their airbrushed selves on digital technology or in social media.' Being on digital technology from a very young age prevents young children from developing emotional intelligence. Anxiety is also increased by our 'quick-fix society' – we want to get rid of unpleasant experiences quickly and don't learn how to sit with discomfort or distress. This undermines our capacity to build resilience.

Lowinger also observes that clients often have some level of anxiety in their family history, so Willow and Heather's story is not unusual.

Lowinger helped Willow see that worrying about something bad happening typically makes the anxiety get worse: 'It's called the snowball thing because when a snowball keeps rolling down the hill it gets bigger and bigger and bigger, and that's what she explained to me.'

Lowinger also taught her a number of strategies for managing her condition. 'The first strategy I think was "put the book on the shelf" because it's the same old book that you read, the same old story, the same old worries.'

Willow continues: 'And also she taught me how to do breathing … Breathe in the air, breathe out the fog, because when your brain is so foggy you can't focus that well, so I'd be able to focus, and I did pretty well in my exams because of that.

'There is this other strategy she taught me about, like, **fight or flight** I'm pretty sure it's called. And so she told me that if you fly from something it will always follow you, and so you've actually got to stand still.

'And she also taught me about the **Worry Bully**. The Worry Bully is … everyone has it, it's this thing inside your head that keeps on telling you worries and worries and worries. And the more you listen to it, the stronger it gets. She told me that the Worry Bully isn't a friend or an enemy, it's just somebody who's there and you can't get rid of him or her.

'She gave me pens and colouring and stuff, and arts and crafts, and we drew our Worry Bully on a paper bag. I kept on looking at the paper bag and I put my worries in the paper bag as if like putting the book back on the shelf, putting the worries back in the Worry Bully.'

So, how does she feel now? 'I feel way better about everything now, I don't really care what people think of me. I've got way more friends and I'm trying to make friendships, my grades are going up now that I am breathing out the fog … so everything is fine.'

What causes anxiety?

'We all experience anxiety to some degree,' says Lowinger, 'because it's part of our primitive survival mechanisms in response to threat in our environment. So our brain responds, the **amygdala*** hijacks our brain and triggers the **fight or flight** reaction, which is a physiological reaction to set our body up to fight or to run away from danger. That response occurs with *perceived* threat as well.'

She adds, 'So we get this surge of adrenaline and cortisol through our bloodstream that primes our body to be able to protect itself, and that physiological reaction is what we experience as anxiety.' (I'll examine the fight or flight reaction again in Chapter 21, when I look at the mind–body connection.)

Lowinger explains that underpinning most anxiety is a fear of uncertainty. 'Commonly in contemporary society it might be fear of being judged negatively, or the perception about what somebody else is thinking of us. So it might be fear of making a mistake – that's performance anxiety – so the fear of being judged negatively and fear of making a mistake, they are subsets of **social anxiety**, and one would be more around interpersonal situations and the other would be more around performance-related situations.

'Other areas where we have uncertainty is in health: what if this symptom or this sensation means that something bad is happening ... But the core fear in these sorts of things is grappling with uncertainty and trying to get certainty when there isn't any certainty, and perfectionism and all of this sort of stuff is trying to get certainty.'

When does anxiety become a mental illness?

So when does it become a problem? 'Where it turns from anxiety into an anxiety disorder is if this threat response occurs to a level that it's providing prolonged fear, suffering and avoidance in a person's life,' says Lowinger.

'Certain individuals with a particular kind of anxious temperament might be more inclined to perceive threat in their environment, and perhaps have a more sensitised amygdala that would be more easily triggered into that fight or flight reaction.'

She adds: 'Really commonly I would find that these individuals have this deeper sense of empathy and a beautiful analytical mind, so deep thinkers, deep feelers. And so when we have a deeper sense of empathy, we are more likely to be aware of emotions in other people. So there's this kindness and compassion that can play out in individuals who experience anxiety more commonly.'

It's estimated that one in five Australians aged 16 to 85 experience a mental illness in any given year, and almost half of us will develop a mental illness within our lifetimes. In Australia, young people have the highest prevalence of any group. It's widely agreed that most mental health conditions stem from a complex mix of factors.

Anxiety is the most common mental health condition in Australia. On average, one in three women and one in five men will experience it, to some degree, at some stage in their life. It often occurs in combination with other common mental illnesses like **depression*** (which I'll look at in the next chapter).

People often respond to their anxious feelings with what Lowinger calls safety behaviours. 'So safety behaviours will be fear-driven behaviours to try and quell that anxiety reaction. So we might conceptualise safety behaviours as things like mind-reading, which is trying to work out what somebody else is thinking about us. It might be trying to avoid or escape from situations, so that's more of a physical behavioural reaction.

'Procrastination is another avoidance type. But then there are other ways that we deal with these unpleasant emotions, and that might be trying to numb the emotions, perhaps looking at drugs and alcohol to try and numb these emotions, or self-harm is a way of numbing some of these challenging emotions, and very sadly we have a horrendous prevalence of suicide in society now, again, in an attempt to get rid of difficult emotions.'

It's natural to be challenged by difficult emotions like anxiety – but when these feelings overwhelm us and interfere with our day-to-day functioning, it's important to seek help.

Mind Strength

Lowinger has compiled a set of strategies to treat anxiety in children and adults, which she calls the Mind Strength Method.

She explains that the Worry Bully 'is what's called externalising the worry ... So we stand up to worry, we replace worry with more practical alternatives such as problem-solving and action planning. We look at what's *in* our control and what's *out of* our control. There is a large toolkit of Mind Strength strategies that help to stand up to those fear-driven thoughts and those fear-driven behaviours.'

If there is a risk that a patient will harm themselves, Lowinger will seek advice from a psychiatrist or GP, but she knows from experience that the strategies she teaches can be very powerful. 'There's a concept that I love to refer to, and it's called distress tolerance. Distress tolerance is one that I think can help so many people, and it's the key message to recognise that it's not weak to feel.'

She explains, 'When we feel that we are weak because we feel, then we push our feelings under the carpet and then we are quicker to choose drugs and an individual may be quicker to take their life because they feel like they're not good enough for experiencing some of these difficult emotions ... We are still raised with these messages, particularly the men in our society, but certainly the women as well, that we have to be stoic, we have to cope.

'But the world is a complex world, filled with stressors that our body wasn't necessarily designed to experience, and naturally we are going to have some emotional consequences to this.'

Whether it's medication or psychological intervention, the bottom line, she says, is for people not to suffer in silence and to seek help when they need it. 'It's just recognising and having the courage to say I need help here, and allowing themselves to acknowledge that it is not a weakness, anxiety is not a weakness, it's nothing to feel ashamed of, it is *such* a prevalent part of our human condition.'

13

DEPRESSION

Conversations with:

Chrissie, who has experienced depression

Professor Ian Hickie, Co-Director of Health and Policy,
Brain and Mind Centre, University of Sydney

Alex Korb, Adjunct Assistant Professor, Psychiatry and Biobehavioral
Sciences, University of California, Los Angeles

Playing the long game: Chrissie's story

I met **Chrissie** when she was in her 30s, but her struggle with **depression*** began when she was a teenager.

'There wasn't a lot of acknowledgement or acceptance of the illness, and also I guess my family was going through a grieving period, as my father had just died, so we just assumed that things were going to be quite awry for a long time. But it just kind of turned into this ongoing, untreated issue for me which would come and go. And in my kind of blind way I stumbled through lots of different attempts to sort it out …

'So I tried various things like meditation, fairly intensive meditation through my 20s, and that was effective. But I wasn't able to keep it up. I tried lots of counselling, which is effective for specific periods of life but it wasn't necessarily treating the state itself of depression, it was more looking at specific problems. But

I think also I was operating on a kind of deliberate blindness as well because I did not want to accept it.'

It was only when she had a child and consulted a psychologist for help with her baby's sleep issues that Chrissie became aware she had depression: 'She actually was the first person who said to me, "You are depressed, you need to have some antidepressant drugs, I'm going to give you an appointment with a psychiatrist." So that was the first time – and I was 35 – that someone had sat down and made a diagnosis.'

She was prescribed antidepressants, which she took for three years. 'But at the same time I flat-lined emotionally,' she says. 'I didn't laugh. I just went through life like a machine. And actually at first it was a real relief, it was great, but then after a while I felt I couldn't keep that up, I wanted to stop, so I gradually came off the medication and I did it really responsibly, just by reducing the dose very gradually.'

Slowly, she got herself off the medication. 'And then I had another big bout. I had some big issues in my life. My partner and I separated and at the same time my mother had a major stroke and, bang, I didn't know what had hit me but I was struggling through it again.'

But Chrissie really wanted others to hear about her experience. 'I guess I'm telling this story bearing in mind that this is a condition that will play out over a long period of time. It's not like a cold where the symptoms are sudden, you feel them, you know that you are subjected to this particular illness.

'It can be so vague. But still every day is a struggle. So I'm just trying to swim and muddle through my life during these periods, and it makes it really tough.'

Later, Chrissie found some relief from her condition by participating in a trial of a promising new treatment: **transcranial direct current stimulation.*** (This is similar to Transcranial Magnetic Stimulation [TMS], which I'll discuss further in Chapter 19.)

What causes depression?

Professor Ian Hickie is Co-Director of Health and Policy at the University of Sydney's Brain and Mind Centre. As he explains, the Australian statistics for depression are nearly on a par with those for **anxiety:*** 'About one in four Australian women and one in six Australian men at some time in their adult life will have an episode of depression that is actually disabling: it interferes with their life and puts them at risk of other complications, such as suicidal behaviour, effects on their relationships, effects on their employment, contribution to poor physical health, and often the development of alcohol and other substance abuse problems.'

He points out that depression is an umbrella term for many different illnesses: 'If you like, depression is like fever, it's something that tells us that something is wrong without necessarily telling us the underlying cause.'

Depression results from a combination of factors, says Hickie: 'We know from twin studies that we did here in Australia and in other countries that it's about 35 per cent genetic ... The rest is due to environment. Environment includes things like stressful life events.' Chrissie's on-again, off-again struggle with depression very much bears this out.

Hickie explains: 'One of the things about the brain that I think has been really important in modern neuroscience is to actually be able to see the extent to which it is changing in structure and in function in relation to the situation in which you find yourself. For example, if you socially withdraw, bits of the brain shrink. Don't ever retire. You need to keep working, you need to interact with people, you need to see smiling faces, you need to interact with the outside world.

'The brain changes its structure constantly in relation to those social inputs, as it does in relation to physical activity, light inputs from the external, from the sky ... dietary factors that are influencing these things all the time. So the brain is a dynamic, reactive structure.

'Now, we have not understood that because we've not been

able to see it in life. What modern neuroscience has delivered us is a way to monitor it in life and see changes.'

Just as there is no one single treatment for fever, Hickie concedes there is no perfect treatment or silver bullet for depression. However, he is co-leader of the Australian Genetics of Depression Study, part of an international scientific collaboration, and he's particularly enthusiastic about how genetic information could help us understand each person's unique pathway to depression.

'When you do come forward with depression, could we pick the right treatment for you the first time around?' he muses. 'At the moment we have trial and error, we have big studies on big populations, on average 50 per cent, 60 per cent of people will improve with this treatment. Unfortunately, often with a lot of the medical treatments, 10 per cent, 20 per cent will be made worse by the treatment, and we have no way of knowing upfront whether you are one of the ones who is likely to benefit or not.'

The study may also help patients deal with the *environmental* side of their depression: 'Actually, because you can control or understand the genetic factors better, you understand the environment better, and you understand what are certain environmental factors.'

As Hickie explains, 'Many of those things are often a *consequence*, they're not the cause. So a lot of marital difficulty, a lot of work difficulty, a lot of financial difficulty is not the cause of the problem, it's the consequence of the problem that, sure, makes it worse over time, but it's not the thing driving the problem. So really understanding the environmental factors better through better characterisation of the genetics and the biology, the two go hand in glove.

'So I'm very enthusiastic.'

Understanding the downward spiral

Alex Korb, a neuroscientist at the University of California, Los Angeles, refers to depression as 'the downward spiral'. He believes that when we're faced with mental health concerns there are small things we can do to change the activity of the brain's neurotransmitter system, which in turn can set us on what he calls an upward spiral out of depression. It may also prevent the development of more significant mental health problems. Korb talks about this in his book *The Upward Spiral: Using Neuroscience to Reverse the Course of Depression, One Small Change at a Time.*

When we met on *All in the Mind*, I said to him, 'Those who have not experienced depression could think that it's feeling sad all the time, but it's not as simple as that, is it?'

Korb's research has confirmed that depression can be different for each individual, he says. 'I've found that over and over again, a specific set of circuits kept coming up, and the way that I like to think about depression is that it's a problem with the way the thinking and feeling and habit and reward circuits in the brain are communicating with and regulating each other, and something is just a little bit off about the dynamics of that communication – it means the person gets sort of stuck in this certain pattern of activity and reactivity that is very hard to get out of.'

One example of the role these brain circuits play in depression is in relation to memory. I've mentioned elsewhere in the book that a key structure for memory is the **hippocampus**,* which is in the **limbic system**, the feeling part of the brain. When people are depressed, says Korb, 'they can't even really remember being happy, and that has to do with the way the hippocampus recalls memory. It works on a principle called state-dependent memory recall that gets the whole context, including the emotional state that you're in, and it makes it easier to connect to memories that are related to that context. So when you're sad and depressed, it's simply easier to remember other times when you're sad and depressed, and all the negative things that led to that state.'

Fortunately, Korb says, it's possible to change the way these circuits act and communicate. Once we have a better understanding of what happens in the brain when we are depressed, there are small life changes we can make to avoid getting stuck in a downward spiral.

One is to understand that human brains are wired to pay more attention to emotional information than plain facts – and some brains have more of a negative bias than others. Korb explains: 'Humans, we're generally optimistic, and that makes negative events much more salient and jarring, because we don't expect them to happen … And obviously some people are more optimistic than others, but the happiness and satisfaction that we feel about life isn't so much based on the good things or the bad things that happen to us, but rather the difference between the good and bad things that happen to us, and what we expected. So if we expect everything to be pretty good, and it's just OK, then we're constantly paying attention and noticing how everything is not as good as we had hoped. Whereas if you just sort of assume everything's going to be OK, and then it's good, then you might be positively surprised.

'We can't always help what we automatically focus on,' Korb says, 'but when we notice that our negative emotions are being driven by the negative things that we're paying attention to, if we just shift our attention towards the positive things in our life, it doesn't make the negative things go away, it just means that they're not front and centre in our attention … And paying attention to the more positive parts of our reality sometimes takes a little bit more effort, because we have these habits of paying attention to the negative that are beneficial in many instances, but if they're weighing on your mood and making everything feel dark and difficult, you can retrain some of those circuits by intentionally guiding your attention to the positive parts of your life.'

Another practical thing we can do to prevent the downward spiral is exercise. Korb explains that 'exercise can modulate the serotonin system and it can modulate the dopamine reward system and the habit circuits in the brain … It's one of the biggest things you can do to start changing the activity and chemistry of your brain for the better.' Other practical strategies include developing better sleep habits, tuning into your body more and ensuring you have social support around you.

In depression we often get stuck when it comes to making decisions, says Korb: 'A lot of times we experience **anxiety*** because we don't make a decision, and then we don't act on that decision, and that gets us sort of stuck in this loop ... We don't actually make a choice and move forward, and the more options you have, the more uncertainty there is, the more you focus on the negatives, the greater your anxiety tends to be and the greater the reactivity in your emotional circuits. Stop trying to make the *best* decision and just try to make a *good enough* decision.'

Intentional decision-making is a great way to start an upward spiral because it engages the **prefrontal cortex,*** which is responsible for goal-directed behaviour. Once a decision is made, the prefrontal cortex helps us to organise ourselves to achieve that goal. Exercising these brain circuits strengthens them.

Another way to counter the downward spiral is simply to express gratitude. 'Sometimes we don't allow ourselves to feel grateful, or to acknowledge that there are good things in our life,' Korb points out. 'And we think that we have to pay attention to something simply because it exists and is true ... but that is, I think, a hard thing for people to accept, and so they continue to focus on things that are true even though they're not helpful.

'So that's one of the things that gets in the way. Another is thinking that you can think your way out of everything ... We like using our brains and our minds to figure out and think through everything, but depression is not always something you can think your way out of.

'In those situations you just have to take action ... and that's one of the main reasons I wrote *The Upward Spiral*: to try and help people understand how the brain works. When we understand these things and why they work and how they affect the brain, sometimes it makes it easier to take action.'

He adds: 'I think the most important element of *The Upward Spiral* is you use it to understand yourself and your brain and how it works, and *your own* tendencies, so that you can start making the choices that are right for *you*.'

14

POST-TRAUMATIC STRESS DISORDER (PTSD)

And other impacts of trauma

Conversations with:

Izidor Ruckel, abandoned Romanian orphan and orphan advocate

Charles Nelson III, Professor of Pediatrics and Neuroscience and Professor of Psychology, Department of Psychiatry, Harvard Medical School, and Professor of Education, Harvard Graduate School of Education

Esther McKay, former police officer and PTSD sufferer and President/Founder, Police Post Trauma Support Group

Professor Sandy McFarlane, Director of Traumatic Stress Studies, University of Adelaide

Bessel van der Kolk, Professor of Psychiatry, Boston University, and Medical Director of the Trauma Research Foundation, Brookline, Massachusetts

Abandoned in an orphanage: Izidor's story

After the overthrow of the Nicolae Ceauşescu dictatorship in 1989, Romania was opened to the world for the first time in decades. When that happened, around 170,000 Romanian orphans were discovered languishing in the harsh conditions of 700 institutions: a generation of children who'd been brought up without care, stimulation or comfort. This prompted the most comprehensive study ever of the effects of **trauma*** from neglect and institutionalisation on children's wellbeing.

Izidor Ruckel, now in his early 40s, was discovered when he was 10 years old within the fortress-like orphanage of the Hospital of the Irrecoverable Children, in the small Romanian town of Sighetu Marmaţiei. Izidor contracted polio and his parents left him in the hospital and never went back. He openly shared his moving story with me, in the hope that exposing his life experience will help to prevent the terrible damage caused to children who are abused and neglected.

His earliest memory is of a woman named Maria, who took better care of him than anyone. She protected him and provided him with things like food and shampoo from her home. 'And then just a few years later she was electrocuted from trying to heat up the water for the kids, and the only reason I found out about it was because the workers knew that she cared for me so much and how much I loved her that they told me.'

He explains, 'Really I didn't have any emotions. I think I was just shocked or stunned to hear that really, and everything just changed from there.' Over the next few years, he and the other children were woken up every morning at 5am, given breakfast and put into a clean room, where all they did was rock backwards and forwards. Some cried, and some would hit themselves and were put into straitjackets to prevent them from hurting themselves.

The children teamed up in an attempt to protect each other from the cruelty of the staff. After a while, Izidor was noticed to be a leader among the children, so the staff would put him in charge of groups of 50 or 60 kids. However, if they were unsatisfied with the way he dealt with the children, he was severely beaten.

At last, after 10 years of being abandoned in this cruel institution, Izidor Ruckel was able to escape. After an American documentary called *20/20* shocked the world by showing emaciated Romanian orphans rocking back and forth in metal cribs, the late award-winning producer John Upton arranged for some of the children to be adopted by American families. Izidor Ruckel was one of them. But his relationship with his adoptive family was rocky because of his extreme emotional vulnerability.

Izidor was very open with me as he reflected on the impact of his institutionalised past. 'My mentality in the orphanage was completely different from the mentality I have [now]. I could not think for myself, I followed orders. Have I been mentally traumatised because of what I went through? Yes, I was.'

He believes that every Romanian orphan has been deeply affected. Most of them desperately want to find out about their past and their birth family, but their biggest struggles are with education and forming healthy relationships.

However, Izidor has done his best to move forward. 'I believe that sometimes God has a plan for you when you don't know it, and sometimes he's going to say, "You know what, all the beatings, all the suffering you go through, I'm still going to keep your mind intact for a reason, so you can raise awareness for other people who are suffering the way you did."'

And he has great hopes for the future. 'One of my dreams is to be able to travel back and forth to Romania. I want to help in Romania. It's not just for orphans, it's for individuals who live on the streets, individuals who struggle so hard that they can barely hold on to life itself. I want to be able to help people like that.'

The effects of neglect and institutionalisation

When the world became aware of the devastating experiences of Romanian children like Izidor Ruckel, neuroscientists saw this as a vital research opportunity. Harvard Medical School professor **Charles Nelson III** led the charge. In 1999 he launched the Bucharest Early Intervention Project, the most significant scientific study ever on the effects of institutionalisation and neglect on children's development. It's a joint collaboration between researchers at Tulane University, the University of Maryland, Boston Children's Hospital and Harvard University.

I spoke to Charles Nelson about going to St Catherine's, the largest orphanage in Bucharest at the time.

'My first visit there was really unbelievably heartbreaking, because if you went through the wards of babies and very young children who they considered to be typically developing, it was heartbreaking to see children who had been basically ignored. There might be 15 children in a room with one caregiver who was really not paying much attention to them, and the first thing that stood out was the large number of children who were rocking or banging their heads or biting themselves.

'And if you went into a room of older kids who could walk, for example, they would walk up to you and sit on your lap, hold your hand, walk away with you, even though they'd never seen you before. And of course that's referred to as indiscriminate behaviour, and that indiscriminate behaviour is sort of a hallmark feature of kids who have been in institutions.' It's an adaptation to their environment, being a good way to get someone to pay attention to them. But Nelson points out that when a stranger acknowledges this behaviour by holding them or picking them up, it indicates that it's OK to go to strangers, and this is not healthy or safe.

The Bucharest Early Intervention Project is a randomised controlled trial – a trial that involves randomly assigning subjects to one of two groups – comparing the effects of foster care and institutionalised care on children who were abandoned shortly after birth. In 2000, the study took random samples of 208 children living in institutions. They placed half of them in good-quality foster care, while the other half remained in institutional care. They then followed the children's physical growth and cognitive, emotional and behavioural development over the next 16 years and are now engaged in follow-up studies.

Nelson described to me some of what they found when they first assessed the children. These children had IQs in the mid-60s to low 70s, which is way below the population average of around 100. 'In fact, in many parts of the world an IQ below 70 allows you to be classified as mentally retarded – an older term now, of course, but the fact is the vast majority of our kids fell into that range. They were very, very delayed in language, they showed many impairments in emotional development, including attachment.'

Nelson explains that brain development is greatly influenced by the conditions we are exposed to after birth. So, for example, many of these institutionalised babies would lie on their backs and stare at a white ceiling for the first year of their life, with no one to talk to them or stimulate them, and the researchers believe this accounts for the deficits and delays observed in them.

In follow-up research, the team used electroencephalography (EEG) to measure the electrical activity in the children's brains. It was dramatically lower in the institutionalised group than in the children who were in foster care. In general, the institutionalised children had IQs 10 to 12 points below those in foster homes, though the age of placement was important. In nearly all cases, the children placed in foster care before the age of two had better outcomes in all aspects of development.

Nelson and his colleagues also looked at how the children made the transition from childhood to adolescence. Nelson told me they were very disconcerted by some of their findings. They used magnetic resonance imaging (MRI) to study the children's brains, and it was found that there was a dramatic reduction in grey matter, the matter within cell bodies, and white matter, the matter within nerve fibres (or axons), which project from the cell and help it communicate with other cells. This reduction, along with reduced electrical activity and a smaller-than-average head size, showed that the children – even those placed in foster care – had lost either brain cells or the connections between them, putting them at risk of falling further behind as they transitioned from adolescence to adulthood.

However, Nelson sees an element of hope offered by the notion of **neuroplasticity.*** Even though the children have fewer brain cells, it could still be possible to build the connections and change the wiring, given an appropriate environment.

After years of studying the effect of institutionalisation on children's development, Charles Nelson's message is clear: the best thing you can do for children who are parentless is place them with a loving foster family as early in their lives as possible.

What is trauma?

Trauma is the response to an event or events that overwhelms the person affected and makes life difficult to cope with. Trauma from a single event – a natural disaster, a serious accident, an incidence of physical or sexual abuse – is usually referred to as **single-incident trauma**. **Vicarious trauma** occurs when someone is exposed to another person's trauma.

Other people experience many repeated traumatic events over time, like ongoing abuse, neglect or violence. This is known as **complex post-traumatic stress disorder (CPTSD)**. Certain jobs carry a high risk of developing **post-traumatic stress disorder (PTSD)***, including the military, firefighters, healthcare professionals, first responders, war correspondents ... and police officers.

Trauma in the police force: Esther's story

Esther McKay has positive memories of her early days as an idealistic young police constable in the tough, male-dominated world of forensic investigation. Her first few years with the New South Wales Police Force were all about learning to do her job well and analyse the crime scenes correctly.

After a while, though, once she felt she'd mastered the basics, she started actually looking around at what she was dealing with. She often found herself in deeply confronting crime and grisly murder scenes, sometimes having to be in a small area alone with a dead body for hours at a time. She'd go home at night full of adrenaline and endlessly mull over the day's events without any opportunity to debrief.

I'll never forget my visit to Esther several years ago. She had been declared medically unfit with PTSD after 17 years on the job. When I spoke to her at her home on the outskirts of Sydney, she'd just returned from a trip to the dentist, where she was being treated for problems with her jaw resulting from anxious teeth-grinding when asleep.

Although there are now better support systems in place within the police force, during those 17 years as an officer Esther felt isolated. Her stress built up to the point where she was really suffering. She had many nightmares and would wake up in the night screaming and crying for help. She told me about one incident where a rack of lamb on her dinner plate reminded her of a postmortem she'd been to, and she burst out crying, pushed the food away and continued to cry all night. Flashback reactions like this one made her sick and shaky, leading to feelings of **depression*** for days afterwards, which she describes as feelings of impending doom.

Another time she attended an accident where a small child had been electrocuted after his sister had thrown a hair dryer in the bath. One morning soon afterwards, when Esther was drying her hair, the electrocution label fell off the dryer onto the bathroom floor. She became paralysed, and she couldn't touch it for days.

Esther was living with the difficult emotions of fear, hypervigilance, anxiety and sometimes numbness, which all affected her daily life and family relationships. After years of stress and exhaustion, she was diagnosed with post-traumatic stress disorder.

She eventually received help by attending counselling and group therapy courses which taught her strategies to manage her symptoms and assist her recovery. I was inspired to learn that since then, she's been extremely active in developing services to support serving and former police officers and their families who suffer from work-based trauma.

What is PTSD?

Professor Sandy McFarlane is the Director of Traumatic Stress Studies at the University of Adelaide and first began studying the effects of trauma after the 1983 Ash Wednesday fires that swept across South Australia and Victoria, devastating the lives of thousands of people.

He told me it's now known that events leading to PTSD occur in approximately 50 per cent of women and over 60 per cent of men. It can form the basis of many more cases of mental illness than we think.

'The core aspects of it,' he explains, 'are firstly that the person in a sense becomes dominated by the past; their mind becomes frequently preoccupied and revisits the traumatic experience. That can happen in a variety of ways, such as nightmares or triggered memories.

'The second part about it is that these people often have a range of quite severe and debilitating anxiety symptoms ... The third part of it is that obviously people have to adapt to living with those sorts of symptoms and they'll do this by doing two things.

'The first thing is to try and avoid either the thoughts or reminders that they relate to the trauma. The second thing is they become quite withdrawn and emotionally numbed and shut off and often these people find it very difficult to project themselves into the future.'

He continues: 'What we're interested in is whether the patterns of activation in the brain in post-traumatic stress disorder are different from those people who don't have post-traumatic stress disorder. In fact what we find are very significant disruptions and one of the most interesting ones is the loss of activation of the area that's involved in expressive speech.' Many people with PTSD report having difficulty at a party for example, engaging in social conversation, or even reading aloud.

The symptom that best differentiates the people who have PTSD from those who don't isn't the nightmares, or the intense traumatic memories, but the disturbed concentration. PTSD sufferers live in a world where they can't distinguish what's relevant from what is not, causing a high degree of anxiety.

Healing trauma

Bessel van der Kolk, Professor of Psychiatry at Boston University, has spent decades studying and treating people haunted by their stressful experiences. He believes that the key to understanding trauma lies in the connection between the brain and the body.

He says standard talking and drug therapies are not effective enough in resolving trauma. His unconventional approach is based on the idea that trauma has nothing to do with cognition, but it is a sign that your body is resetting itself and reinterpreting the world as a more frightening and dangerous place. He says self-regulation – calming your brain down and focusing on the present – is central, as is self-awareness.

By investigating the brains of PTSD patients using medical imaging, Van der Kolk and his colleagues have found that deeply traumatic experiences literally rearrange the brain's wiring. When people remember their trauma, they lose touch with their current environment and go into survival mode. They keep reacting to normal situations as if they are in danger, resulting in a sense of helplessness and an inability to move when faced with stressful experiences.

This is why child abuse so often causes lasting psychological damage. Normally our brains are wired to expect that we will be comforted and consoled by others, but in situations of child abuse, the source of comfort is also the source of terror.

According to Van der Kolk, the main function of the brain is to take care of the body, so when people are traumatised and feel threatened or enraged, all these things are experienced at a bodily level.

When a person constantly feels in danger, they shut down and can't always identify what is safe and what is dangerous. Van der Kolk believes this explains why some people seem to go from one trauma to another, often deliberately putting themselves back in harm's way. They may even feel more alive in dangerous situations.

Van der Kolk treats many children who have been traumatised by abuse, and he takes issue with common therapeutic approaches such as exposure therapy or cognitive behavioural therapy, where patients are often asked to speak about their upsetting experiences. *His* approach begins with helping children feel safe, then he asks them to do physical activities like jumping on trampolines or walking over balance beams. This helps them tune in to their bodies and their sensations, feeling what pleasure is and what sensations they want to avoid.

Van der Kolk's trauma treatment approach is considered unconventional and lacking in scientific rigour by some mainstream psychiatrists. However, he supports any therapy that pays close attention to our physiological state. These include activities like yoga that change our **heart rate variability (HRV)**, a measure of the variation in time between each heartbeat. According to Van der Kolk, HRV plays a crucial role in our response to trauma. It's controlled by the **autonomic nervous system (ANS)**, which regulates our heart rate, blood pressure, breathing and digestion, and controls our **fight or flight** and **relaxation responses** (which I'll look at in Chapter 21).

Healthy people typically have a high HRV, which means their pulse fluctuates rapidly in response to external stimuli, and they can moderate their emotions by controlling their breathing, in order to stay calm and focused on the present moment. When people become stuck in their traumatic past, taking short, rapid breaths out of worry or anxiety, their HRV is poor, and changes in breathing take much longer to affect emotion.

In his research, Van der Kolk saw better results with yoga in people with chronic PTSD than with any medication that had been studied. He finds this quite exciting.

I wondered how Van der Kolk has been affected by listening to disturbing stories of trauma over many years.

'It's exhilarating,' he told me. 'Basically everybody I treat is an awesome person, is a person who when I sit with them I go like, "Boy, if I'd had your history, if what has happened to you would have happened to me, I have no idea how I would have survived."'

15

BORDERLINE PERSONALITY DISORDER (BPD)

Conversations with:

Gabby, diagnosed with borderline personality disorder

Eliza, Gabby's partner and carer

Sathya Rao, Adjunct Clinical Associate Professor, Monash University, Melbourne, and Executive Clinical Director of Spectrum Personality Disorder Service for Victoria

Dr Maria Naso, psychiatrist

Riding the roller-coaster

Gabby was on an emotional roller-coaster, feeling empty and needy. Whenever she lashed out in anger, she'd regret it and say sorry repeatedly. Her partner, **Eliza**, felt like she was walking on eggshells, always fearful of arousing Gabby's intense emotions.

Gabby had started seeing counsellors and psychologists at around 12 years of age. She'd been diagnosed with a range of conditions, including ADHD and bipolar disorder, but her symptoms didn't seem to match any typical cases. So, for a long time no one knew what was going on with her and she was just considered angry. But then, in 2011, Gabby was diagnosed with **borderline personality disorder*** by two psychologists and a GP.

Borderline personality disorder was named during the early days of psychoanalysis, when it was thought to be at the border between **neurosis** (a milder stress disorder) and **psychosis*** (a more serious disorder involving loss of touch with reality). It's no longer seen that way, but the term has stuck, and sufferers still tend to feel stigmatised.

When Gabby was diagnosed with borderline personality disorder she didn't take it very well and thought the psychologists had it wrong, until Eliza made her aware of how her behaviour was impacting on her. Though Gabby was never physically violent towards her, Eliza said, 'I am scared to come home.' She told Gabby that she had to accept the diagnosis and get help.

People with borderline personality disorder often self-harm, and it is estimated that 10 per cent of them will die by suicide. Gabby says, 'I'm not a self-harmer in the traditional sense, but I will become physically aggressive towards myself, so I'll hit myself or I will hit a wall, just basic anger, just uncontrollable rage. I can come across as very angry, very vicious, when that's not normally who I am. I'm just erratic.'

Eliza recalls that when Gabby's condition was at its most extreme it was an unpredictable and trying time, and she never knew how Gabby might react to the simplest of statements. '[A request like] "Could you pop those dishes away?" one day would get a normal response, but the next day would trigger quite an emotional response, which then led to self-loathing for overreacting. It just spiralled.'

Although the causes of borderline personality disorder still aren't fully understood, it's thought that genetic influences and trauma can play a part. There is a history of mental illness on both sides of Gabby's family, and she was a victim of sexual abuse between the ages of seven and twelve. This sexual abuse was not acknowledged by her family, which exacerbated her trauma and distress. By the time she was in her early 20s, she felt her reactions to many everyday situations were not normal.

'I bounced from friendship to friendship for a long time, and relationships – this is my longest [she has been with Eliza for over 10 years]. For someone with borderline personality disorder, trust is really hard because you don't feel like anyone understands you, because you don't even understand yourself. Being trapped in your mind can sometimes be very scary, and that's what it does.'

Because Gabby's condition was trauma related, she had some specific trauma therapy, and then she underwent **dialectical behaviour therapy**, which she says 'absolutely changed my life'. Dialectical behaviour therapy has proven to be one of the most effective treatments for borderline personality disorder. It helps sufferers move away from destructive coping behaviours to a more fulfilling life, by teaching them skills such as **mindfulness,*** how to tolerate distress by distracting and self-soothing, how to regulate their emotions, and how to manage relationships.

Gabby did a 20-week group program of dialectical behaviour therapy which taught her strategies such as naming emotions, which is not easy for someone with borderline personality disorder. 'It's just this anger, this aggression,' Gabby says, 'you don't know where it's coming from. Mine comes from a lot of guilt and fear. I live in constant fear. I constantly had my head looking behind my shoulder, you know, what's coming next? But being able to name that emotion and then being able to work with it and think, what's this emotion telling me, and how can I react to this emotion, is really helpful.'

The program also helped Gabby talk to Eliza about her feelings, which in turn gave Eliza a better understanding of Gabby's issues. Eliza also did a support program for carers and families, which improved their relationship even more.

Gabby wants to communicate to others with borderline personality disorder that 'I understand the pain; I know it's scary, but you can live a great life, you really can. If you step through the door and get the help you need, I promise you, putting the work in is worth every step of it.'

Of her partner, Gabby says, 'There is so much I admire about Eliza. Her willingness to stand by me when a lot of people wouldn't have ... is incredible. She is incredible. I admire her persistence, I admire the fact that she has always made me feel loved, and she's my carer but she is also my sidekick, she is the one I lean on. There's not enough words to describe what she means to me and I'm just very lucky.'

And Eliza can now easily say what she values about Gabby: 'The capacity to never give in, the passion that Gabby has for life and that she is starting to put that passion into herself, that's something that's very admirable. Just not quitting: she's not going to give in, not going to play the victim card. She has such a capacity for love and understanding. And I think a lot of what has gone on for her in this journey is that she wants to speak up and help other people. BPD is still so stigmatised and taboo; even within the medical professions it's just so misunderstood. Gabby wants to help get the information out there about it, to show people that it can be lived with and lived through. I'm just so proud of how far she's come and how she continues to work and push herself to live the best life and not waste any of it.'

BPD: The psychiatrist's view

It's estimated that 10 to 13 per cent of the world's population have a **personality disorder**. Such disorders develop when your character traits and behaviour cause you intense distress and strongly deviate from social expectations.

Borderline personality disorder is among the most common, causing deep emotional pain for approximately 1 per cent of Australians. In clinical populations women are over-represented, but in the community both genders are equally represented. Unfortunately, many men don't get recognised with BPD and often end up in drug and alcohol or prison settings. Symptoms usually appear in the teenage years or young adulthood. Having BPD is not the person's own fault; it is a disorder of the brain and the mind. For too long it's been misunderstood, but now, with appropriate treatment, up to 80 per cent of BPD sufferers can dramatically reduce their symptoms and learn to live normal lives.

Associate Professor Sathya Rao from Monash University is a passionate advocate for people with this diagnosis. Rao describes people with BPD as having intense and unstable emotions, out of proportion to the situation. He says people have described BPD as like having third-degree burns over their whole bodies. It's so mentally painful it can feel like lacking emotional skin.

Impulsive behaviours, such as overspending, gambling and binge eating are common. Many people with BPD get involved in situations with unanticipated consequences. They struggle with their identity and often have unstable relationships. They frequently use drugs, alcohol or self-harming behaviours in an attempt to fill the void. 'Non-suicidal self-injury is a signature symptom of BPD. Patients self-injure to regulate painful emotions – feeling suicidal can act as a coping mechanism to help them survive.' Rao adds that '10 per cent of people [with BPD] can die by suicide. Having a diagnosis of BPD may reduce a person's life span by 20 years.'

Rao explains that the causes of BPD still aren't completely understood, but it's known that there are biological, genetic and psychosocial influences. **Trauma*** is an important aspect of the disorder: patients are wired to feel, think and behave in a different way because their brains have been sculpted by their earlier experiences. However, Rao stresses that trauma is a risk factor for BPD, but not a prerequisite for the condition.

BPD has also had the reputation of being a very difficult, if not impossible, condition to treat effectively, because people with BPD can be uncooperative in treatment, often ambivalent and confused. Rao adds, 'People with BPD may have difficult behaviours, but it is the disorder that causes it. It is misunderstood as deliberate and manipulative.'

Unfortunately says Rao, this has meant that the mental health workforce is not well trained in this area, and can be the biggest stigmatisers of the condition.

Rao says people with BPD are often given prescription medications, but this is not appropriate: there is no medication that specifically treats the disorder. Instead, when people with BPD are given evidence-based treatments like **dialectical behaviour therapy** – the treatment Gabby has undertaken – 60 to 80 per cent of them go into remission within one or two years, and others make a full recovery and go on to lead productive lives.

There's now evidence that this doesn't need to be long-term therapy. Short-term interventions can be just as effective.

Dr Maria Naso is a psychiatrist practising in South Australia. Sometimes she's called on as an expert psychiatric witness in court cases involving people with a diagnosis of BPD. She has often seen them being treated very poorly, showing that the stigmatisation continues. When she was training as a psychiatrist, there was a sense that people with BPD were manipulating the system and were untreatable.

This perception must change. People with this diagnosis need just as much help as other patients who present with conditions such as **schizophrenia,** * **bipolar disorder** * and **depression.** * BPD is a legitimate and complex psychiatric illness that needs to be adequately understood, resourced and treated, without placing the blame on the patients.

16

HEARING VOICES
Auditory hallucinations

Conversations with:

'**Margaret**', voice-hearer

Professor John McGrath, Conjoint Professor,
Queensland Brain Institute

Professor David Copolov, Pro Vice-Chancellor, Professor of Psychiatry
and Honorary Professor of Physiology, Monash University, Melbourne

Professor Susan Rossell, cognitive neuropsychologist and
Professorial Research Fellow, Centre for Mental Health, Swinburne
University of Technology, Melbourne

Professor Flavie Waters, Research Professor, School of Psychological
Science, University of Western Australia

Ron Coleman, voice-hearer and owner of Working to Recovery

Indigo Daya, voice-hearer and mental health advocate

Dr Rufus May, clinical psychologist

Will Hall, voice-hearer, therapist and group facilitator

Amanda Waegeli, voice-hearer

The voice of God: The story of 'Margaret'

Margaret (not her real name) was diagnosed with **schizophrenia***
at the age of 21. She remembers her first **voice-hearing***
experience as very real, profound and frightening.

She would hear three predominant and distinct male voices, and she called them Dennis, Mark and Adrian, which are the names of people she has met in her life. She felt they represented parts of herself that needed relationship and connection. The voices were often quite rude and at times would make her feel very agitated and aggressive in her internal commentary back to them. But sometimes they would give her advice.

Later, her voice-hearing took on a religious flavour: she'd hear frightening demonic voices, angelic voices and even the voice of God. She became paranoid in public, adrenaline was constantly pumping through her body and her moods were erratic. She told me that it was like being a mother and having children around you while you are trying to cook or speak to someone else and they are constantly interrupting you and disrupting your concentration. She thought her voices were definitely part of her, yet they also felt not of her, like they were some sort of intervention.

Margaret believes her voice-hearing experiences reflect **trauma*** from earlier in her life. She remembers being sexually abused, and in early adulthood she had an unpleasant sexual experience. Shortly after the onset of her psychosis, that voice very much represented the abuser. Throughout her voice-hearing experiences, there have been links to her life history, and she feels the voice-hearing is encouraging her to put together pieces of her past and reconcile those parts of herself that have been lost or disconnected in some way.

Margaret still hears voices, having deliberately chosen to stay on a low dose of medication, because she'd prefer to have a good level of functioning, but she's developed strategies to manage her condition. She says good morning to her voices and embraces them in a kind and respectful relationship that she feels they deserve. She doesn't fight against them any more and now believes that a person can change the relationship they have with their voices and find meaning in the experience. Sometimes she even says thank you to her voices for giving her a different perspective on things.

When Margaret and I chatted, she had a message that she really wanted to share with me – that even though a person may hear voices or have received a diagnosis of a mental health condition, it doesn't need to define them. She believes that such a person can still contribute to society and have a meaningful life, and there is no reason why anyone can't reach their goals and fulfil their potential.

All about auditory hallucinations

Hearing voices is the most common form of hallucination, particularly in those with **schizophrenia*** or **psychosis.*** But recent research is beginning to show that these hallucinations can occur as a part of a range of other mental disorders, and even in otherwise mentally healthy people.

Professor John McGrath from the Queensland Brain Institute says the scientific understanding of hallucinations has been dramatically transformed in the last 10 years. He and his colleagues have done a large amount of research on the link between hallucinations and mental illness. In collaboration with researchers around the world, they've looked at voice-hearing across the general population. In a group of 30,000 people, they found that about 1 in 20 had had at least one hallucination across their lifetime. It seems these experiences are more common than was once thought.

McGrath says this finding also counters the myth that if you hear voices, you are schizophrenic, or on the way to becoming so. There is a proportion of individuals who are mentally healthy, yet report hearing voices. For this group the hallucinations are usually not very frequent or distressing, but they do occur.

These initial findings led McGrath and his fellow researchers to explore the link between hearing voices and other mental illness. They found a link between voice-hearing and common mental health issues such as **depression*** and **anxiety*** – which makes sense, because if you are anxious, hypervigilant and easily startled, you readily misinterpret outside cues, and that may trigger a hallucination.

It's important to note that if you have auditory hallucinations but are otherwise healthy, you don't need to see them as a warning sign for mental illness, because the hallucinations tend to fade away. McGrath advises that people who've heard voices two or three times don't need to seek help – but if the voices persist or are distressing then that is a concern.

McGrath's research shows that we need to reshape the way hallucinations are considered in the general landscape of mental illness. It could help scientists discover why some people recover and others go on to develop more serious disorders such as schizophrenia.

What happens in the brain during auditory hallucinations?

David Copolov is Professor of Psychiatry at Monash University, and for 20 years he was the Director of the Mental Health Research Institute of Victoria. He and his colleagues have undertaken a series of studies that show particular regions of the brain are spontaneously activated during hallucinations. One is the **temporal lobe**, which is associated with the processing of normal sounds, so it's as if the brain is being tricked into believing these voices are actually occurring. The other regions activated are those associated with memory, such as the **hippocampus**.* So, Copolov believes that the auditory hallucinations are likely to be the result of having to make sense of reactivated auditory memories, resulting in the false perception of external or internal voices.

The voices experienced by people with mental illness such as schizophrenia often have derogatory connotations, 'saying' things that imply that the hearer is a bad person. Margaret's hallucinations follow this pattern.

In some cases the hallucinations urge the person to carry out undesirable acts, including acts that might be harmful to themselves or to others; these are known as **command hallucinations**. Around 70 per cent of people with schizophrenia have auditory hallucinations, and about half of those people have command hallucinations.

Importantly, says Copolov, not all voice-hearing involves negative experiences. In fact, about a third of the people he and his team spoke to said that they would miss the voices if they disappeared. A considerable number of them said the voices were helpful and guiding. As mentioned, Margaret has learnt to tolerate hers. On the whole, though, voices associated with psychiatric illness cause the hearer distress, and living with these intruders can significantly disrupt people's lives.

Professor Susan Rossell is a cognitive neuropsychologist and and Professorial Research Fellow at the Centre for Mental Health at Swinburne University of Technology in Melbourne. When she first started in the field of voice-hearing over 20 years ago, there was massive controversy over whether these experiences were real, or only in people's imaginations.

Now, she says, brain imaging technology has been able to show that when someone is having a voice-hearing experience, exactly the same brain regions are active as when we listen to someone speaking to us face to face. The main brain regions are **Wernicke's area** in the temporal lobe, which is involved in *comprehending* speech and **Broca's area** in the frontal lobe, involved in *producing* speech.

Why some people have the voice-hearing experience and others don't is the million-dollar question, says Rossell, but science is getting a little closer to having some answers. One idea she and her colleagues are looking into is that those who hear voices are having difficulty monitoring their own thoughts. We all have inner speech, which activates the brain regions described earlier, and we can choose to listen to our thoughts or not, but those who have auditory hallucinations may have an inability to tune out their internal dialogue.

The Hearing Voices Network

The Hearing Voices Network was established in the early 1990s by Marius Romme, a Dutch psychiatrist, after one of his patients,

Patsy Hage, persuaded him that the distressing voices plaguing her were real, not a figment of her imagination. Together, doctor and patient went on a Dutch television chat show and were overwhelmed by the audience response. After further research, Romme became convinced that validation of the voice-hearer's experience and support from others with similar experiences can play a crucial role in people's recovery.

The Hearing Voices Network approach is powerful, because when people share their experience with others they start to learn new coping strategies. They include encouraging people to negotiate with their voices and suggesting that they listen to them at a particular time of the day then ask the voices to leave them alone so they can get on with their lives. Another strategy is voice dialogue, where another voice-hearer actually has a conversation with the person's voices. This often allows the voice-hearer to gain more clarity about the role of their voices in their life. What will work for one person may not work for another – but the beauty of the Hearing Voices Network is that people share their insights into different approaches.

Professor Flavie Waters, from the School of Psychological Science at the University of Western Australia, believes the Hearing Voices Network offers many people invaluable support and a sense of empowerment. 'Voice-hearers want to have more decision-making in determining the type of treatment they are getting. The Hearing Voices Networks have contributed significantly towards a first-hand perspective of what it's like to have hallucinations and hear voices, and to a better understanding of the help options available to those who need it,' she says.

The emphasis of the networks is often less on medication and more on psychological treatment, open dialogue and working with the voices so they are perceived as meaningful. 'So they incorporate it as part of their personal narrative, their personal experience, and this is seen as being really greatly therapeutic.'

'Psychotic and proud': Ron's story

Scotsman **Ron Coleman** is a larger-than-life character with a cheeky sense of humour. He has a tattoo on his arm that says *Psychotic and proud,* and one of his favourite T-shirts has the words *I hear voices and they don't like you* printed on the front.

He told me about his first experience of hearing voices that aren't there.

'I'd just broken my hip playing rugby and after I came out of hospital I went back to work and I was sitting in my office one day and I heard a voice behind me say, "You've done it wrong." And I looked behind me and there was nobody there. From then on it got worse, and worse, and worse, I wasn't hearing one voice, I was hearing two, then three.'

'And eventually I was hearing six, and that led me into hospital. The voices are so intense and they always pick on the bits of us that are our weaknesses. They leave us feeling as if we are worthless and hopeless and we become victims of that, really.'

He told me, 'I don't hear them inside my head, I hear them outside. It's almost like I'm hearing you now.'

His voices were distinctly different. The most distracting was that of a priest who had abused him as a child, telling Ron that the abuse was his own fault, that Ron had led the priest into sin and he deserved to burn in hell. He also heard the voice of his first partner, who had died by suicide. Her voice told him to kill himself so that they could be together again.

'And that became a real hard part of my life, and eventually I just couldn't cope with it any more and I broke down totally.'

He spent six and a half years out of the next 10 in and out of hospital. 'At the end of 10 years, I wasn't hearing six voices, I was hearing seven voices. So actually I came out of the system in a sense worse than I went in, in terms of the voice-hearing experience. Because the voices never changed, they were always there.'

Finally there was a turning point. 'I was in Manchester, I was in hospital and I heard a support worker who suggested that I went to this new group that was just forming in Britain called the

Hearing Voices Network. Even at the very first meeting it made sense to me. A woman said to me, "Do you hear voices?" And I said, "Yes", and she said, "They're real."

'And as I thought about that it became clear that for years I'd been told they weren't real and I had to ignore them and therefore I was powerless, I couldn't do anything about it. And now somebody was saying to me, "This experience is real so you have to do something about it, there's no point waiting for other people to do something for you." And then I started going more and more to the self-help group, to the Hearing Voices group.'

Ron points out that 'it's about understanding why the voices are there. It's very easy to take a position that they're there because of some chemical imbalance and yet … about 70 per cent of voice-hearers can relate their voices directly to what has happened in their lives.'

The Hearing Voices Network changed everything for Ron: 'Through the group and through working on issues in my own life I was able to take control over the voices and experience and get on with my life. Married, got children, done all the things that I really wanted to do but was unable to do through that period of pain when the voices had control.'

Since then he's been passionate about raising awareness and understanding of voice-hearing. Though he's had to step back more recently due to other health issues, he's been influential in the development of the Hearing Voices Network in the United Kingdom, and has worked with his wife, Karen, to support people in their recovery from **psychosis.***

'The Judge': Indigo's story

Indigo Daya has been in and out of mental institutions, and reached 130 kilograms due to the side-effects of antipsychotic medication.

'I experienced a number of different types of trauma in my childhood,' Indigo explains. 'I won't go into lots of graphic details, because that can be distressing and it's not really necessary, but I experienced neglect at quite a young age and at times physical abuse. Then there was a period where I was bullied and beaten up by local kids for being Jewish.

'Anyhow, life became difficult because of that and a range of other things at home. I ran away from home and very sadly on that first day I was abducted. And so there was a two-week period where I experienced sexual abuse and other not very great things. So that's probably enough said, that's enough to experience, really.'

Indigo was 30 years old when she first heard voices. She calls one of her most prominent voices 'the Judge'. She doesn't hear it through her ears, as many voice-hearers do; it's more of a thought or entity that she hears inside her mind. The Judge has always been extremely critical of her, using rude language and telling her that she's evil.

It was when she realised that the voices she hears are to do with the trauma of being sexually abused that a breakthrough finally came.

'The connection for me was when I went to hear a speaker, so it was someone else who had been diagnosed with schizophrenia who had been in mental health services, and his name is Ron Coleman, he's from Scotland. And I heard Ron speak about his own madness, about his own trauma and how it wasn't until he understood the trauma experiences that he could make sense of his voices and begin to heal. And that was really new to me but it was incredibly inspirational.

'And after I heard Ron, it kind of set me on a path of listening to other consumers and survivors who had been through similar experiences. And they kept sharing similar stories, that their healing had come about by understanding that the mental illness, the madness, made sense in the context of what had happened in their lives. And so that set me on the path to going, "Well, I need to think about why this has happened to me."'

Now, with lots of support and conversations with others who've been through a similar experience, Indigo realises that the voices are her way of trying to make sense of the awful things that happened to her. It represented her shame about the abuse, for which she blamed herself, but now she understands that none of it was her fault. And while she still occasionally hears the voice of the Judge, she's not frightened by it any more; it's just pointing out to her that she needs to forgive herself.

Indigo is well on the path to recovery. After going through intense withdrawal symptoms, she no longer takes antipsychotic medication. She is now a strong advocate for the growing Hearing Voices Network.

Indigo feels that the stigma around voice-hearing can often stand in the way of healing. To her, the perception that people who hear voices are dangerous and that voice-hearing is linked with madness and violence is an offensive and unhelpful misconception, because the majority of people who hear voices have been victims of trauma and abuse.

Other treatment approaches

According to David Copolov, treatment for voice-hearing could involve a relationship with a therapist, talking about the voices with others, and antipsychotic medication for hallucinations and similar symptoms. However, in about 25 per cent of cases involving **psychosis**,* medication doesn't significantly reduce the voice-hearing. In these cases, psychological therapies can help people cope with their voices.

At the same time, other practitioners are developing new approaches to treatment – with promising results.

Engaging the voices

Dr Rufus May has been a clinical psychologist in the United Kingdom for more than 20 years. His interest in holistic approaches to healing from mental illness and voice-hearing began as a result of his own **psychosis*** and recovery in his late teens.

He's passionate about using different approaches to help people deal with their voice-hearing if it's disturbing to them. He runs therapy groups where the goal is not to get rid of the voices, but to establish a new relationship with them. One approach – similar to the Hearing Voices Network's voice dialogues – is to engage with the voices, and even to engage one voice with another.

His technique comes from couples counselling, and the idea is that we all have different parts of ourselves that we need to manage and negotiate with, and this can be done with voices too. So, for example, he asks the voice-hearer if he can converse with their different voices and question them or negotiate with them directly. He might ask one of the voices, 'How long have you been in Tom's life?', or 'What do you want from Tom?', or 'Why are you angry with Tom?' May tries to find some common ground between the voices, and in a sense acts as a mediator.

He treated one woman whom he wasn't able to see very often, so he asked one of her voices to send him a Facebook message. The voice replied that it wanted its own Facebook account. After that, a Facebook group was set up so voices from all around the world could talk to each other. May says this is a place where people can come to terms with their voices and collaborate with others. It's a really different way of relating to our minds, where we embrace the diverse voices within us.

Will Hall also has personal experience of voice-hearing, and is now a therapist in the United States. He too uses an approach of engaging inner voices. His technique acknowledges that many of us have a critical inner voice that puts us down and focuses on negative parts of ourselves. We don't all experience that as an external entity, but it's similar to the experience voice-hearers have.

Hall coaches people using role play, to help them talk back to parts of themselves, to be assertive and even to shout at the voices to tell them not to be mean. The challenge at the same time is to be open to listening to the potentially valuable or helpful messages the voices may be conveying to them.

Hall suggests that this is a method that could be helpful to everyone when they are feeling overwhelmed by their inner critic.

Compassion-focused therapy

In 2018 I took part in a compassion-focused therapy workshop, as part of a conference held by the International Society for Psychological and Social Approaches to Psychosis. The workshop was run by UK psychologists and researchers Charlie Heriot-Maitland and Eleanor Longden. Longden tells of her own compelling experience of hearing voices in her widely downloaded TED talk 'The Voices in My Head'. In her moving account of her long journey back to mental health, she reveals that it was through learning to listen to her voices that she was able to survive.

In the workshop, Heriot-Maitland led us through a series of exercises based on neuroscience to create a sense of safety. We were then encouraged to explore the different sides of ourselves – the angry self, the anxious self, the sad self etc – and try to cultivate a more *compassionate* self. I was surprised by how challenging it was to really tap into a feeling of compassion for myself; it felt much more difficult than having compassion for others.

Compassion-focused therapy is still new, but has been successfully applied to problems like eating disorders, **depression*** and severe self-criticism. But Heriot-Maitland has also seen some promising results for voice-hearers. More scientific research using randomised controlled trials is the very important next step.

* * *

During this workshop I had a powerful encounter with **Amanda Waegeli** about her own voice-hearing experience. She started hearing voices when she was a teenager, after she'd been abused as a child. Despite this she managed to get through high school, but 12 years later the voices returned, as she was in a difficult marriage. Then she was diagnosed with post-partum psychosis (a serious mental illness that follows childbirth).

That was when she entered the psychiatric system. She was heavily medicated and had electroconvulsive therapy, but after eight years the voice-hearing was the same, if not worse.

Then she linked up with the Hearing Voices Network, and through sharing her experience with other voice-hearers she learnt more about the identity of her voices, and how to manage them. She sees it as like being in a relationship you can't get out of: there's tension, but you've got to find ways to meet in the middle.

Amanda still hears voices – in fact, she told me she was hearing them during our conversation, because she was anxious, and the voices were telling her that she was going to say all the wrong things. She was trying to slow down and look at me and just focus on what we were saying.

Amanda is also a songwriter, and she gave me a CD of some of her music. She described the challenge she had in recording one of her songs, 'Amanda's Dance', which is about her voice-hearing experience. Her voices weren't too happy about it, and every time she went to sing, they were screaming in her ears, making it very difficult to stay in tune. Eventually she managed to get it recorded, which pleased her, because the song is very special to her.

17

INTERGENERATIONAL TRAUMA

Conversations with:

Professor Pat Dudgeon, psychologist and Poche Research Fellow,
School of Indigenous Studies, University of Western Australia

Jacqueline McGowan-Jones, Commissioner for Children
and Young People, Western Australia

Rowena Cox, suicide prevention advocate

Dr Tracy Westerman AM, Clinical Psychologist,
Indigenous Psychological Services

An introduction to intergenerational trauma

Intergenerational trauma* is a relatively new field of study, first
recognised in 1966 when Canadian psychiatrist Vivian M Rakoff
and her colleagues found high rates of psychological distress among
the children of Holocaust survivors. In 1988 they published a
study showing that grandchildren of Holocaust survivors were
300 per cent more likely than the rest of the population to be
referred for psychiatric care. More recent research has identified
intergenerational trauma among other groups such as the indigenous
populations of North America and Australia.

There is now some evidence that children can be affected by
the **trauma*** experienced by their parents before they are born,

or possibly even before they are conceived. Research into the mechanisms by which trauma is passed through generations is complex and still evolving, but scientific studies are identifying epigenetic mechanisms that may be in play.

Epigenetics* is the study of the effects that environment and behaviour have on our genes. Trauma can cause changes in a person's DNA (deoxyribonucleic acid), their unique genetic code. These so-called *epigenetic* changes don't damage the genes, but can switch them on or off, altering how they function. Research is showing, however, that, unlike other genetic changes, epigenetic changes are reversible, making it possible to restore healthy functioning with adequate social support and a change of environment.

Suicide in Indigenous Australian communities

Even in the 21st century, Indigenous Australians sit on the wrong side of a harsh racial divide, and are affected by a history of suffering dating back to the earliest days of European colonisation. They have one of the highest incarceration rates in the world, and it's estimated that the Indigenous suicide rate is at least double those within the general population. Staggeringly, over 8 in 10 of these deaths occur between the ages of 15 and 44. In children aged 5 to 14, suicide rates for Indigenous boys and girls were 9 and 7 times as high, respectively, as those of non-Indigenous children. (I'll talk more about why young people are particularly prone to mental illness in Chapter 19.)

At the heart of this crisis is trauma – not only that occurring in the present, but also the trauma that has been passed down through the generations. This intergenerational trauma is the result of colonisation, violence, loss of land and culture, and policies such as the forced removal of children.

It's now recognised that intergenerational trauma is common within Australian Aboriginal and Torres Strait Islander communities, particularly among the descendants of the Stolen Generations.

Australian Indigenous leaders acknowledge that many of their people are dealing with the devastating impacts of this condition.

Professor Pat Dudgeon is a Bardi woman from the Kimberley. A leading figure in suicide prevention, she's a psychologist and a Poche Research Fellow at the School of Indigenous Studies at the University of Western Australia. She explained to me that the elevated Australian Indigenous suicide rate follows a pattern seen within indigenous populations across the world who've been colonised by a dominant power. The issues that have led to this suicide crisis include relationship and family breakdown, bereavement, **depression,*** **anxiety,*** substance abuse and involvement in the justice system. The grief, loss and racial discrimination experienced by all Indigenous Australians, as well as a widespread lack of access to safe and culturally appropriate support services, likewise have an impact.

Jacqueline McGowan-Jones, currently the Commissioner for Children and Young People in Western Australia, was, when I spoke to her, the CEO of Thirrili National Indigenous Postvention Service. She says there are several misconceptions about the causes of Indigenous youth suicide.

People point to drugs, alcohol and violent relationships as major causes, when they are really the symptoms. The cause of just about every case of Indigenous suicide, she says, is trauma. It may be related to being taken into out-of-home care for example, or it may be intergenerational trauma. Because Indigenous people have an oral culture, they can unknowingly retraumatise future generations when talking about some of the struggles and tragedies that have affected their people over the many generations since European colonisation.

Aboriginal leaders are working together to reduce the devastating suicide rate in their communities. One such initiative, led by Professor Dudgeon, was the Aboriginal and Torres Strait Islander Suicide Prevention Evaluation Project. It assessed the effectiveness of existing suicide prevention programs and delivered its final report in 2016.

Dudgeon told me that two significant issues came out of that report. One was that Aboriginal and Torres Strait Islander people need to be in charge of identifying the problems, and to be part of the solutions as well. The other important finding from the report is that programs have to be culturally appropriate. They must recognise, acknowledge and strengthen the culture of Aboriginal and Torres Strait Islander people.

The year after the report was released, the Centre of Best Practice in Aboriginal and Torres Strait Islander Suicide Prevention was established to further investigate effective approaches to this issue.

Fostering hope through footy

Rowena Cox is a Djaru woman from Halls Creek in the Kimberley, in Western Australia, where she has been a community liaison officer. She lost her son to suicide nearly a decade ago. Now she draws on her own life experience to help and support others.

She told how she has a deep understanding of the trauma people are experiencing in her community because she's lived through it herself. Now she wants to educate those in her community so they can identify the problems they're going through themselves, and get the support and care they need.

Her son was a talented and enthusiastic football player. She remembers that when she asked him to do chores around the house, he never listened, but when his friends suggested a game of footy he'd be out the door. After he passed away, his mates and cousins suggested Rowena form a football team in his memory.

The team went to a festival in Fitzroy Crossing in the Kimberley, and Rowena says it was overwhelmingly emotional to see these young men proudly playing in memory of her son. They didn't play with their heads, they played with their hearts, and when they came home undefeated it uplifted the whole community.

Before every game she takes them to, she reminds them that they are part of a team that is promoting suicide prevention in their communities. 'And I'm so proud, because there are other young men of the team who have also lost their loved ones, and just seeing them play, it's overwhelming and beautiful, really.

'I am passionate about promoting suicide prevention,' says Rowena. 'It just breaks my heart every time I hear of young people passing away; even if they are not from my community, you know, I tend to shed a tear, because I know exactly what their parents are going through and how their parents are feeling.

'So I'll be doing this for a long time to come yet. I just really want to do it to encourage other people to support suicide prevention. If *I* can get help and be where I am now, anyone can do it.'

Culturally appropriate treatment

Tracy Westerman is a Nyamal woman from the Pilbara region in Western Australia. She's the Director of Indigenous Psychological Services and an adjunct professor in clinical psychology. She's recognised as a world leader in Aboriginal mental health, and in 2018 was named Western Australia's Australian of the Year.

Westerman has aspired to be a psychologist since childhood, and is passionate in her belief that access to culturally appropriate psychological services is key to suicide prevention among Indigenous people. 'I find that big systems don't move very quickly unless you provide them with irrefutable evidence of cultural difference,' she told me. Up until now, she says, we have often confused risk factors for suicide with causes. For example, alcohol abuse, poverty and the effects of colonisation are considered to be risk factors but by themselves do not explain suicidal behaviour. Causes include depression and trauma. She says if you eliminate the *cause*, you eliminate the end result.

Part of her research has involved developing a type of science of culture – using assessment tools to explore with Aboriginal people their experience of suicide to see how it compares to suicide in mainstream Australian society.

Westerman says that Aboriginal people make meaning of the world in a very different way from non-Indigenous people. Normal Indigenous cultural behaviours are often seen by mainstream psychologists as indications of mental illness. For example, it's common when someone has passed away for Aboriginal people to experience a spiritual 'visit' from that person. From a Western medical perspective, it's easy to diagnose that as a sign of mental illness. Australian Aboriginal people are diagnosed with **psychosis*** and **schizophrenia*** at about 17 times the rate of non-Indigenous Australians, despite the fact that these illnesses are very rare. This is why Westerman feels that understanding, treating and healing Indigenous people are tasks best left to *Indigenous* psychologists.

In her view, Indigenous mental health needs to become a national priority. 'I think there are so many gaps that have not been addressed, and I'm hopeful that they will be, because the gaps are actually very clear.'

18

DISSOCIATIVE IDENTITY DISORDER (DID)

Conversations with:

Kallena Kucers, DID survivor

Professor Warwick Middleton, Director, Trauma and Dissociation
Unit, Belmont Private Hospital, Brisbane

Dr Cathy Kezelman, President, Blue Knot Foundation

Rhonda Macken, writer, composer, music educator and DID survivor

'I' or 'we'?: Kallena's story

I really didn't know what to expect before my conversation with
Kallena Kucers. Unfortunately I couldn't meet her face to face,
as we were in separate states, so we did an audio interview.

She was immediately warm, calm and thoughtful. In fact, in a
strange way, she put me at ease.

When I spoke to Kallena she was 52. In her mid-30s, she'd
been diagnosed with **dissociative identity disorder (DID).***

'Before that I had been quite successful in study, career. I had a
family. I thought my life was going really well, to be honest – or
at least the parts of myself who were living my life at that stage
thought things were going quite well.'

They were parts of herself – often called **alters*** – that had been
formed to help her manage living everyday life in a way that wasn't

possible for the other 'parts' that knew about the traumatic things that had happened to her. The feelings, memories and physical sensations connected with her past pain had been separated out.

Kallena lived with so many different identities that she said it was impossible for her to describe them all. There were the ones that turned out to be quite fundamental to who she is, and another layer of selves that helped her practically from day to day. 'There were also very dominant ones that knew a lot about the particular traumatic things I'd experienced as a child and also then as a young adult. Then there were also multitudes and multitudes of parts that simply were … I prefer to call them just bits, just fragments of memory, short events, feelings, a lot of body memories and so on.

'The ones that really lived the life most, they were focused on either being able to do particular kinds of things or express particular kinds of feelings. Like, for instance, there was one who just liked to have fun and who was good at sport because we thought sport was fun, things like going skiing and so on. So she was good at those kinds of things and loved them and had a lot of fun, and was good at teasing people and being very social and so on, whereas others were very introverted.

'There was another one who was very good academically, who really had no connection with the body at all, who wouldn't have been able to ski for the life of her, but was academically very good and successful.'

Kallena's parts had different physiological features and abilities too. For instance, some needed glasses and some didn't. Some could speak English and others couldn't.

These identities had varying levels of awareness of each other, and how they each felt. Some had no awareness that any others existed at all, and believed they were the only part, or alter, in Kallena's life.

Largely Kallena was able to function reasonably well, because none of her parts ever blatantly showed themselves to other people; the whole reason for being like this was to hide what had happened to Kallena in her past. However, there were times in her life when her parts became more obvious, when she was very unsettled or under a lot of stress.

In her late teenage years, Kallena was still in an abusive environment and at the time she was diagnosed with **bipolar disorder*** because of her extreme mood swings: sometimes she seemed depressed, sometimes manic. The mental health workers weren't aware that she was actually switching between a shut-down, traumatised alter and an extroverted, fun-loving part.

Most people with DID have experienced severe **trauma*** in childhood and dissociate from the violence and trauma in their past as a way of handling day-to-day living. The condition, in a way, provides an effective protective mechanism.

Kallena certainly has trauma within her family background: 'It was, looking back on it now, an extremely dysfunctional family, with a lot of things that happened in it that really shouldn't happen to any child at all. Starting from when I was absolutely an infant, I was treated in ways that were extremely hurtful and painful and terrifying. A lot of times I probably believed I was going to die, which is I think at the core of why a lot of these separations happen – you really believe you're going to die.'

Many of Kallena's memories of this frightening time are still quite clear to her – although for much of her life there were 'parts' of her that had no idea that any of it had happened.

For years, the complex dissociative condition Kallena developed allowed her to lead a relatively normal life. It wasn't until she started a family and her young children reached the same age when the abuse had begun in her own childhood that she became very distressed and needed help. She then decided to seek help so she could understand where these feelings were coming from. This was when she started long-term psychotherapy.

'And the way that works really is it's about relationships,' Kallena explains, 'because the trauma originally happens in relationships. So in that therapeutic relationship someone begins to feel safe enough for all the different parts to actually begin to show themselves and to begin to develop some kind of relationship with one consistent other person, and then gradually begin to feel this person as safe enough to express more and more of what they know, what they feel.'

She explained that the therapist respected all the parts equally, so they began to cooperate and communicate with each other. After years of work, she was able to attain a type of co-consciousness, in which all parts are present and aware of what's happening, and tolerate the overwhelming feelings contained in the separate parts.

Ultimately, the parts just blended together, and Kallena now has one whole self. The healing process felt like a real team effort!

I noticed as Kallena was explaining these different parts of herself, she often used the pronoun 'we' rather than 'I'. Her identities were obviously still very present to her, even though 15 years of therapy had allowed her to integrate into one identity. But I wondered, now that all Kallena's parts were together, how it felt for her emotionally.

Her response was so surprising to me. She said at first it was terrible, because it was so different. For example, the days felt very long. Previously each part had no concept of how long a whole day took, because they were constantly switching between each other. They also constantly changed who was dominant and made the decisions.

So, is Kallena a happier person now that she is integrated into a single identity?

'Yes, I think I do feel better in myself overall now, yes, but it's a controversial question, because this issue of whether it's something to aim for, or whether it's right or even desirable to get to a point of no longer having parts ... it's something that's not right for everyone. And yes, I do find myself being a lot more content than I ever would have before.

'On the other hand, I do feel I actually miss my parts, certainly at times, as a sense of comfort to having others inside you who you know and can communicate with in this sense of never being alone. I do miss that, there's times I really miss them ... But overall, no, I'm glad I'm like this now.'

My interaction with Kallena left me in great admiration of her. She was so generous, open and insightful about her experience of living with DID. She is obviously motivated to raise the public's awareness about this disorder, to help others who've had similar experiences, and to break down the stigmas and the harmful stereotypes surrounding it.

What is dissociative identity disorder?

Dissociation in psychology is a mental process involving disconnection from one's thoughts, memories, feelings, actions and identity. It is a feature of several disorders, including dissociative identity disorder. People with these disorders involuntarily escape from reality, which can cause them problems in daily life.

Professor Warwick Middleton is Director of the Trauma and Dissociation Unit at Belmont Private Hospital in Brisbane and past president of the International Society for the Study of Trauma and Dissociation. He describes DID as a disruption in the processes of consciousness, memory, identity and perception, processes that are usually integrated.

'A person who is highly dissociative may switch into identity states where they feel that they are of a different age, a different sex, living in another time and place or responding to a different environment. It's a spectrum,' he says.

'And characteristically people who have dissociative identity disorder don't usually come to a health professional saying, "I think I've got DID"; they probably come because they hear voices, they have a history of self-harm, they may self-medicate using alcohol, drugs etc.'

This is because they are usually not fully aware of their condition: 'Characteristically, people with DID switch between different identity states, none of which has the full range of memory and the full range of affect that an integrated personality that's non-dissociative has. So one way to look at it is, yes, you might have multiple identities and name them, and some of these may almost have delusions of separateness, in that they in that state do not believe that they are connected to the other states. I've literally had people say things like "We are not multiple" or "Look, I'm not multiple but I think some of the others are."'

Unfortunately, there is a great deal of fear and stigma around DID, fuelled by the stereotypes portrayed in films and popular culture (which I'll look at next).

Middleton says research shows that the vast majority of DID patients have a history of sexual, physical and emotional abuse or deprivation from an early age, often continuing for a long period of time. So often the child hasn't developed a healthy sense of identity and has no way of escaping from the people who are their primary caregivers but also their primary abusers. This creates a double bind, and the only way to cope is to find a way of compartmentalising the trauma, while at the same time eking out some sort of relationship with the person or people who are causing that trauma. Occasionally a very disturbed, traumatic relationship can go on for decades, and the person is still being sexually abused by the same perpetrator 50 years later. Sometimes the victim is literally imprisoned, though psychological imprisonment can have just as deep an impact.

Estimates on the prevalence of DID vary widely. It may affect between 0.01 and 15 per cent of people and is more commonly diagnosed in women than men. It occurs more often in areas that have experienced large-scale trauma, such as war or natural disaster.

But, as Kallena's story shows, there is hope. 'The overall finding,' says Middleton, 'is that in the vast majority of cases, when a patient with DID enters into some regular structured therapy with somebody who has sound boundaries and a functional understanding of trauma, things improve.'

He explains, 'It's not like you've lost anything, you've just taken away an artificial partition; it's like one of those office plans where you can have movable walls and move progressively towards an open-plan office where everyone can see everyone else.'

Sensationalised on the silver screen

In 1976, during my studies in psychology at Sydney University, I was bowled over – and, to be frank, quite horrified – by a television movie that claimed to depict the experience of living with dissociative identity disorder.

Sybil, starring Sally Field and Joanne Woodward, was based on the 1973 book by Flora Rheta Schreiber. Set in the 1950s, it claimed to be the true story of a shy young graduate student who developed **multiple personality disorder**, now known as dissociative identity disorder, as a result of psychological trauma she had experienced as a child. Schreiber gave her subject, Shirley Ardell Mason, the pseudonym 'Sybil Dorsett' to protect her privacy.

In both the book and film, Sybil suffers cruel, sadistic physical and sexual abuse at the hands of her mother, Hattie. She develops 16 personalities, and her psychiatrist, Dr Cornelia B Wilbur, encourages these different selves to communicate in order to reveal information about Sybil's early life. After intensive, dramatic therapy, Sybil eventually becomes integrated into one mentally healthy personality.

I was fascinated by this topic long before Sybil's story was popularised. I recall, one night as a child, watching TV over my parents' shoulders as they viewed another film based on a woman's development of multiple personalities. *The Three Faces of Eve*, released in 1957, won Joanne Woodward an Academy Award for her role as Eve. The film was based on an article and then a book of the same name by American psychiatrists Corbett H Thigpen and Hervey C Cleckley. They told of their therapeutic work with their patient Christine Costner Sizemore, who was said to have multiple personality disorder. The book sold hundreds of thousands of copies and the film caused a sensation in 1950s American popular culture.

Multiple personalities: The controversy

By the early 1980s, after *Sybil* had captured the public's attention, the number of reported cases of multiple personality disorder skyrocketed. It's estimated that by the early 1990s, more than 20,000 cases had been diagnosed in the United States. (A similar rise in cases had occurred after the release of the book and film of *The Three Faces of Eve* in 1957.)

A great deal of controversy ensued about the legitimacy of this disorder, and the role played by early trauma and abuse. Scepticism and debate surrounded the concept of recovering repressed memories of trauma through psychotherapy, and therapists were accused of planting false memories in the minds of their patients. Serious challenges emerged for the legal system when people accused of serious crimes admitted during therapy came forward proclaiming their innocence.

Having been shocked and mesmerised by *Sybil* in the 1970s, my attention was drawn to the release of a book in 2011 by American journalist Debbie Nathan, provocatively titled *Sybil Exposed*. She contends that the story of Sybil was in fact an elaborate hoax, created by the therapist, the patient and the writer.

Despite the ongoing controversy, multiple personality disorder has long been recognised by psychiatrists, and has been in the *Diagnostic and Statistical Manual for Mental Disorders* since it was first published in 1952. In 1994, the fourth edition of the manual, **DSM-IV**, changed the name to dissociative identity disorder to reflect a better understanding of the condition, in which there is a fragmentation of identity rather than a growth of separate personalities. In the most recent version, **DSM-5,*** further changes have been made to clarify that a person who has **alter*** identities and some amnesia is no longer automatically considered to have a mental health disorder. They can only be diagnosed with DID if they experience distress or difficulties as a result.

* * *

The condition of DID was once again on my mind as personal accounts of childhood trauma emerged from the Royal Commission into Institutional Responses to Child Sexual Abuse, held between 2013 and 2017. The deep and devastating impact such trauma has throughout people's lives became publicly and painfully clear.

In the final year of the royal commission came the release of yet another film dealing with the disorder. US horror movie *Split* tells the frightening story of three young girls who are kidnapped and terrified by a disturbed man with DID who has 23 personalities. Yet again we saw this disorder sensationalised by the popular media.

The release of *Split* was strongly criticised by the Blue Knot Foundation, a national organisation that supports the millions of adults in Australia who've experienced complex trauma. President **Dr Cathy Kezelman** told me: 'When you take a mental health issue like dissociative identity disorder, which is a very serious mental health issue related to having been usually profoundly abused or traumatised in childhood, and perpetuate myths and stereotypes that people so affected are vicious psychopaths, it really puts back the progress that we've made by decades.'

Kezelman, who has undergone her own struggle with mental illness, describes how the portrayal of the main character in the film *Split* differs from the people they see diagnosed with this disorder.

'Rather than being violent, they often have a high degree of self-loathing and they often turn that in on themselves ... And it can be incredibly confusing and frightening for people as they struggle with losing time, struggle with different voices speaking who they don't know, and struggle to make sense of their own world.'

Kezelman's words made me reflect on the damage and pain that many of the sensationalist stories about dissociative disorders have caused for far too long. Surely it's not justifiable to produce entertainment that further stigmatises mental illness – especially illness triggered by childhood trauma and abuse.

Hope Street: Rhonda's story

The book that had landed on my desk drew my attention immediately.

It was around two years after my moving encounter with Kallena, and the title intrigued me. *Hope Street – A Memoir of Multiple Personalities: Creating Selves to Survive.*

I noticed that the book had been self-published, and that the author, **Rhonda Macken**, had worked for many years as a music educator and producer at the ABC. I devoured the contents as quickly as I could.

Afterwards I was in shock. Firstly, because of the extent of the abuse Rhonda had suffered for many years at the hands of her father. Secondly, by the astoundingly intricate network of alters Rhonda had created. These identities all seemed to work together to form an extremely effective protective mechanism, to allow her to function in the world, during and after the horrific abuse she endured throughout her childhood.

I must admit that I had some nervousness about interviewing her. Could this extreme story really be true? But I decided to go ahead with it and its truth became very apparent later and was validated by her brother and sister.

Rhonda was accompanied by her husband, Bruce, when we met. She was warm and quietly spoken and openly expressed her nervousness.

Just before we started our conversation, Rhonda paused. There was something she wanted to know from the outset: did I think she was mad?

This caught me off guard initially, but I gathered my thoughts and told her about my previous conversation with Kallena. I explained how much it had affected me and the insights it had given me. So, no – I *didn't* think she was mad.

Rhonda, now in her mid-70s, has been through decades of therapy. As memories came up, writing became an enormously important part of her recovery. Eventually they led to her well-written and engaging memoir.

Rhonda was raped by her father from the time she was two years old, and even became pregnant to him twice. The children didn't survive as Rhonda miscarried as a result of further abuse by her father. He had returned from the Second World War with **post-traumatic stress disorder*** and had regular psychotic episodes. 'He just wasn't there, there was no conscience. So he was very violent. My life was threatened often, in many ways, a lot with weaponry, because he was in New Guinea, with Japanese weaponry, so there were axes and cleavers and even knives.'

Even so, he was the only one of her two parents she could connect to. Her mother had the best of intentions, but she needed to keep a veil of silence around the reality of their family life, so her behaviour towards her daughter was always very cold. She was desperate just to protect herself. Having to rely on her father for affection put Rhonda into a terrible emotional bind.

She began therapy in her 40s, but it was only after starting with her second therapist that she realised she had multiple different alters, who started to come to the therapy sessions.

How many of these alters were there? I asked.

'Well, we got to about 200 and then I stopped counting. There were fantasy parents that I created instead of my parents, there were children, there were teenagers, and there were other strange imaginative people, like Wicked Witch and Persephone and Mephistopheles, and all kinds of people.' There was also an alter she called Master Planner, and his right-hand man, The Jostler. They were all part of a complex jigsaw of protection she'd created.

This protective mechanism was so effective that in her early adulthood she still thought that she'd had a happy childhood. She met and married Bruce, and it was only after her first child was born that she realised something was wrong. She became very anxious and depressed and was referred to a psychiatrist for help.

Rhonda's therapy with her last therapist, Libby, went on for 18 years. She appreciated the fact that Libby was never intrusive but would engage the different personalities in the sessions as if they were real people. Whenever one of them arrived at a session, Libby would greet them. She would recognise their body movements and the way they talked, allowing each of them to tell their story.

The therapeutic process had to be very gradual, so Rhonda could adjust to the full horror of her early experience. At times it was quite traumatic as Rhonda relived very painful events.

Rhonda now says she is fully integrated into one identity but there is grief associated with letting go of some of her alters. She has found her 'self', the child that was born, and believes that now, in this real world she is just Rhonda, but she still acknowledges other parts of her – they are sometimes helpful to her, but no longer dysfunctional.

Rhonda became visibly upset when she told me that her father died when she was 27. 'My father was the human soul on this earth that I connected with, because in his sane, rational times he loved animals and he also was a gardener and I loved the garden, and also I was connected to him to such an extent – I think it's called Stockholm syndrome – where he became almost part of my skin, and to separate from that was a huge process.'

Later, she and her therapist Libby decided that it would be good for Rhonda to scatter his ashes in the ocean. Rhonda recalled that one of her alters, Mary, then appeared and tried to swim out after the ashes, not wanting to let go of her father.

I had to admit to Rhonda that I found her memoir very confronting because the violence and trauma she'd suffered were so extreme.

'I didn't want to put all of those things in the book,' she told me. But Libby pushed for the book to be written, and told her that unless she described real instances of the trauma it would be very hard for readers to understand her condition, and why her imagination needed to create a different world.

As Rhonda puts it: 'When the "what is" defies sanity then the "what is not" becomes sane. So, when the visible world we live in is unsafe, a different reality must be invented. One must create a "what is not" as a tangible alternative to the unsafe "what is". This becomes a credible place in which to escape. The "what is" then becomes real and therefore safe.'

Amazingly, Rhonda has emerged from years of therapy as a healthy, well-adjusted person, surrounded by her loving husband, Bruce, and their children and grandchildren. She's so thankful for her two children. 'I'm very, very grateful that I was given the opportunity to know about love and unconditional love, and I'm very grateful that they are not damaged. They are functioning human beings, they are loving and they have a lot of empathy, they're very creative, and they are wonderful parents. And I'd also like to say thank you to my husband, who I love dearly.'

After this Rhonda and I reunited with Bruce, who'd been waiting outside. As they were leaving, Rhonda stopped and looked me in the eye. She had one more request: she wanted to give me a hug for hearing and believing her story.

I was deeply moved.

The raw, revealing conversation I had with Rhonda, and my earlier encounter with Kallena, reminded me afresh of just how extraordinary the brain is. It can adapt to extreme situations and create super-efficient protective mechanisms – allowing people to endure otherwise unendurable experiences, and simply get on with their lives.

I will never forget either of these amazing women.

Part IV

NEW PATHWAYS TO
HEALING

19

REVOLUTIONARY NEW TREATMENTS FOR MENTAL ILLNESS

Conversations with:

Mike Slade, Professor of Mental Health Recovery and Social Inclusion, University of Nottingham

Peter Raymond, Parkinson's disease sufferer and deep brain stimulation patient

Peter Silburn, Honorary Professor, Queensland Brain Institute, and Co-Director, Asia-Pacific Centre for Neuromodulation

'Matthew', who has experienced clinical depression and transcranial magnetic stimulation

Dr Ted Cassidy, psychiatrist, and Executive Chair and co-founder of Monarch Mental Health Group (previously TMS Clinics Australia)

Professor Paul Fitzgerald, Head of the School of Medicine and Psychology, Australian National University, and co-founder of Monarch Mental Health Group (previously TMS Clinics Australia)

John Elder Robison, author and advocate with autism, transcranial magnetic stimulation patient

'Sok', post-traumatic stress disorder sufferer and neurofeedback patient

Mirjana Askovic, senior psychologist and coordinator of neurofeedback, New South Wales Service for the Treatment and Rehabilitation of Torture and Trauma Survivors (STARTTS)

Dr Roger Gurr, Clinical Director, headspace Early Psychosis Youth Service; Associate Professor, School of Medicine, Western Sydney University

Dr Ben Sessa, psychiatrist, and co-founder and Head of Psychedelic Medicine, Awakn Life Sciences Corp

Professor David Nutt, Edmond J Safra Professor of Neuropsychopharmacology and Head of the Centre for Psychedelic Research, Imperial College London

Rick Doblin, founder and Executive Director, Multidisciplinary Association for Psychedelic Studies (MAPS)

Dr David Erritzoe, Clinical Senior Lecturer and Consultant Psychiatrist, Centre for Psychedelic Research, Imperial College London

The 'recovery model'

The **recovery model*** is just one of the many innovative approaches to mental health treatment that have emerged over the past few decades. As our knowledge of neuroscience increases with the assistance of rapidly evolving technology, societal awareness of what contributes to mental wellbeing is also expanding at a rapid rate. While pharmaceutical medication can be invaluable for many people suffering from mental illness, the medical approach is not always the most effective, and can be complicated by undesirable side-effects.

The recovery model says that taking people's personal life experience into account is just as important for recovery as traditional medical treatment. It grew out of a grassroots consumer movement in the United States in the late 1980s and early 1990s, and it rings true for many people because it concentrates on you as a person, rather than just a set of symptoms.

According to **Mike Slade,** Professor of Mental Health Recovery and Social Inclusion at the University of Nottingham: 'It's kind of changing some of the givens. So the old story was this metaphor of you have a problem, you get an expert to fix it and then you're better and move on. So you get treatment and then you get into life. The new paradigm is you get into life, *that's* step one of recovery. And that often increases motivation and skills at getting better. So you might have, for example, help and support with particular symptoms, but that's positioned as towards the end of pursuing your dreams and goals.

'So the big paradigm shift is that, for example, clinical expertise moves from being centre stage to being one resource among others. Some people talk about services being on *tap*, not on *top*. So it's a resource to offer to people in their journey of recovery, with a modest understanding that for some people actually it may not be the right kind of help, they may need help from their natural supports in their lives or from their community or other spaces.'

Slade very much sees this as reflecting changes within society generally. 'We would expect to make decisions about ourselves and not be told how to live our life, we would expect to have an experience of agency and empowerment and the right to make choices, we would see this as core human rights in terms of our role as a citizen in our country. And it is to some extent bringing that social justice perspective into the mental health system which is the big paradigm shift.'

One of Slade's most frequently cited research papers is a gold-standard overview of the world literature about what people say has been helpful on their own recovery journey. From that research, Slade and his colleagues developed the **CHIME** framework. CHIME is an acronym standing for Connectedness, Hope, Identity, Meaning and Empowerment.

He says that the liberating thing from these findings is that these individuals don't talk about getting rid of symptoms, highlighting that for some people their journey of recovery may not be about eliminating mental health problems. It may be about, for example, reconnecting with their family of origin or their spiritual side, hope that there's a possibility of being better now or in the future, or having a sense of identity that is beyond being a mental health patient.

Deep brain stimulation

A treatment known as **deep brain stimulation*** has recently provided encouraging results. Patients who have undergone the therapy for Parkinson's disease acclaim it as 'brilliant' and describe how subsequent to the treatment they sleep better, walk without tripping or shaking, and stand taller.

Professor Peter Silburn is passionate about the potential for deep brain stimulation (DBS) to relieve not only the symptoms of neurological conditions like Parkinson's disease, but also some mental health disorders.

'It's extraordinary,' he says. 'We're starting to really get into deep-space exploration in the human, not outer-space exploration.'

Silburn is a neurologist and a world expert in the treatment of Parkinson's disease. Together with his colleague at the University of Queensland, Associate Professor Terry Coyne, he has performed more than 800 DBS procedures over the past 20 years, relieving patients of many of the debilitating symptoms of Parkinson's disease, epilepsy and Tourette's syndrome.

The procedure involves drilling holes into the skull and inserting tiny electrodes into the very deepest parts of the brain. These electrodes can stimulate and modulate the neurons and their networks. Silburn describes this as like placing electrodes into the telephone exchange, influencing whole cities in the human brain. The electrodes transmit an electric current that can be turned up and down to adjust how the particular part of the brain operates.

Guided by brain-scanning technology, the probes are inserted while the patient is anaesthetised. Then the patient is woken up in the middle of the operation, 'And we can talk to people and we can ask them questions, we can ask them to think about moving, we can move limbs and watch the neurones fire off. So we are in a position now of not just helping but we can actually watch human thought. And I think that's the exciting part for the future as the technology gets better.'

If someone has a movement disorder, or a bad tremor, once the surgeons see that has improved, they know they have the probes on the right spot. The effect of the stimulation can be seen in front of their eyes during the operation.

They then replace the probe with one that permanently stimulates the brain and sew the patient's skull back up. Next they use that device to start programming the brain. The settings are regularly monitored, and the patient is even able to adjust them with a remote-control device, if needed.

Silburn is confident that this therapy has the potential to treat psychiatric disorders. 'People are exploring for **dementias*** now, and certainly psychiatric disorders such as obsessive-compulsive disorder, phantom limb pain. People are very closely exploring **depression**,* addiction … It really has opened up a great new important way of controlling bad things that happen to humans.' He notes that deep brain stimulation is reversible, and no brain structures are damaged in the process.

'Our future focus on deep brain stimulation should be to identify new conditions that can be helped and what circuits go wrong in the human brain, and that requires a lot of mathematicians, physicists, computational neuroscientists. But we can actually obtain that data now, and I think that's what we should be looking at.'

Transcranial magnetic stimulation (TMS)

'I was a really high achiever at school and then at uni,' says **Matthew** (not his real name). Then in his third year at university, everything started to go wrong. 'I was profoundly sad, miserable, despairing; lots of tears, panic attacks … It was officially diagnosed as clinical depression by my GP, who put me on one of the first standard medications, which didn't work. We tried another one, but my condition deteriorated. Then I was forwarded on to a psychiatrist, and from there things got even worse …

'Over 11 years with my psychiatrist, we tried everything, every class of medication … yet I remained dangerously depressed.

Eventually my mother said, "OK, I've heard about this thing called TMS." It was something I was vaguely aware of, and I was absolutely desperate, so I said, "All right, let's give it a go." And I'm very glad that I did.'

Matthew is among the 30 to 40 per cent of people with clinical depression who don't respond to antidepressant medication. He was treated by psychiatrist **Dr Ted Cassidy**, co-founder of TMS Clinics Australia (now Monarch Mental Health Group), where they use **transcranial magnetic stimulation (TMS)**.* It's a non-invasive method of passing a magnetic pulse through the skin to a focused part of the brain.

Professor Paul Fitzgerald is the Head of the School of Medicine and Psychology at the Australian National University and co-founder of Monarch Mental Health Group. He explains how the process works: 'So when somebody undergoes a TMS session, a coil is placed on the head and an electrical current passes through that coil and generates a magnetic field. There is no contact to the electrical current with the individual, but the magnetic field passes into their body ... the field generated by a TMS coil is very focused. And when you apply a very focused changing magnetic field to something that conducts electricity, it will induce an electrical current, and it does that in the nerves of our brain because they are electrical conductors.

'So when you apply a TMS pulse that is strong enough it will make the nerves in the brain fire ... what we are doing is we are repeatedly applying those pulses to try to change brain activity, because the neurons in our brain will adapt if they are repeatedly firing. There are ways we can use TMS to *increase* brain activity, and there are ways we can use TMS to selectively *decrease* brain activity. And so we tried to target particular brain areas where we think there's value in altering brain activity for treatment of a specific disorder.'

Fitzgerald reports that evidence over 25 years shows that TMS is a safe and effective treatment for people with persistent ongoing depression who have tried two or more antidepressant medications without great success. There's no evidence of longer-term adverse effects, and side-effects are minimal.

So, how did it feel to have TMS treatment? I asked Matthew.

'It's not pleasant, it is not painless. I do believe it's different for all people, and at first I liken it to really bad brain freeze,' he responded. 'And it's only for the four seconds when the machine is switched on … But for me the pain was so significant that I couldn't even handle the maximum dose for the first week, we had to just gradually go up and I'd just be gritting my teeth, I needed to squeeze something just to get through it. But you get used to it.'

He adds: 'Now, despite me saying all this, it's worth it. Not a little bit but a decent amount of pain at the start, compared to all the side-effects on medication, and especially that this is something that works … It's not a miracle but this treatment has had an effect, a significant effect … For me it was a game changer, and I really think it could be for a lot of other people.'

Cassidy says that those who respond well to TMS will feel the benefit more quickly than most do with antidepressant medication. Usually you need between 12 and 20 sessions over a two- to four-week period.

For Matthew it took a bit longer: he noticed an effect after about 40 sessions. Every now and again his depression will relapse, and he knows he can go back to Cassidy for further treatment.

Matthew is the first to admit that he's been lucky his parents helped him pay for his treatment, because at the time it was expensive: a standard course could be up to $5000. In November 2021, though, TMS was made available on the Medicare Benefits Schedule for people with depression who haven't responded to other treatments.

Cassidy and Fitzgerald are both excited about the as yet untapped potential of TMS. Cassidy points out: 'There's some really interesting work with neuroplasticity … And what is clear is that when you give TMS you actually improve the mouldability of the brain. So that really means you can actually learn a bit better. And so when people say maybe TMS makes you smarter, that's what they are talking about, they are talking about the mouldability of the brain increasing at the time of TMS.'

Fitzgerald explained to me the focus of his ongoing TMS research. 'We are still doing work in the area of depression around trying to develop a much more personalised form of therapy … So that's really a major focus in the depression area, as well as trying to produce treatment schedules that are much more efficient. We've been doing research recently trying to show that we can get patients better in maybe two weeks instead of four to six weeks, because that obviously has significant impacts on individuals' lives.

'And then the other area in which we are fairly heavily engaged is exploring the use of TMS for other applications, so we are conducting research in obsessive-compulsive disorder, in **Alzheimer's disease**,* but also in a number of other conditions like trying to treat the voices, or auditory hallucinations, heard by people who have **schizophrenia**,* looking at whether we can reduce some of the symptoms of post-traumatic stress disorder and even try to improve some elements of brain function in people with **autism**.'*

A different world: John's story, continued

In Chapter 7 I introduced **John Elder Robison**, who became known as 'the weird guy' during his career as a rock music engineer in the 1970s and 1980s, and only discovered he was on the **autism*** spectrum around the age of 40. In 2008, researchers from Harvard Medical School approached him about participating in a TMS experiment.

'When I heard that these scientists had this new therapy that might turn on the ability to read emotions in other people, I thought, "Could that really be true? And if it was true, what would it mean for me?"'

John described to me what happened on his way home from one of his early TMS sessions.

'I turned on my iPod because I would play old recordings of concerts that I had worked with back in the '70s and the early '80s, and I turned on the music, and even now just to remember it is just so overwhelming. I turned it on and it wasn't like I was listening to a stereo in a car, it was like I was back there in a nightclub in Boston and it was 1977 or 1982, and I was standing there at the edge of the stage and I was watching the musicians and listening, and it was just so alive. It was just unbelievable, the intensity of it.'

John became quite tearful as he spoke to me. 'And the thing that was most remarkable was that I felt all the emotion of the music that I never could before … And it was just so overwhelming, it was just like magic. And I got home and I wrote them an email and I said, "Boy, that's some powerful mojo you've got in that machine."'

This experience also made him a bit sad, because he realised how many other emotional experiences he must have missed out on. In the next series of transcranial magnetic stimulations he did acquire the ability to read people's emotions and social cues better but also found out that not everything was happy and joyous – there was a lot of pain, trickery and cheating in the world too. 'When they first proposed this to me they said they hoped that this could help autistic people see emotional cues in other people and I thought, "Well, maybe that's why I'm sad and I'm isolated and alone, because I can't see all these messages." I thought immediately of good messages …

'And of course over the next series of stimulations I did acquire the ability to read these messages from people, and they weren't happy and sweet and beautiful and lovely, they were fearful and jealous and anxious and worried and scared and angry … And you know, the pain that I had … I realised that my autism was a protective shield. I hadn't seen that for all those years, and it just about killed me, receiving that pain from others.'

The fallout didn't end there. 'My marriage collapsed, my business almost collapsed. And the thing that's really sad is that it was because suddenly I saw the world and all the people around me differently, and I rejected things that I had accepted all those years.'

It was six years after the treatment that I spoke to John, and fortunately a lot had changed. He told me: 'I'm married again, I have a good wife and family life, my business is better than it was before, my ability to engage with folks like you ... I could never have come to you and had a conversation like this 10 years ago ... But boy, it was a really rough ride to get there.'

Even though the effects of TMS are ephemeral, John feels that once you experience a change in perception like he did, even if the effects don't last, your life is permanently changed in a positive way.

However, he has a few words of caution for those on the spectrum: 'TMS is not and never will be a cure for autism. Autism is a neurological difference that will always be with us. But it is a tool that may remediate things that disable us and make us suffer. I say that in recognition of the really sad fact that the suicide rate for autistic adults is nine times the suicide rate for folks who aren't autistic. And if TMS could head off some isolated person's suicide, just like an antidepressant could do, I think that's a tool that we have a duty to make available to those who wish to use it.'

Neurofeedback

Sok (not her real name) is a quietly spoken Cambodian refugee who survived the Pol Pot regime and arrived in Australian in 1983. She suffers extreme symptoms of **post-traumatic stress disorder*** and **depression.*** When I spoke to her, she told me she still carries a lot of fear from living through the Pol Pot regime, as well as from traumatic childhood experiences. It has had a major impact on her health, and she was unable to continue working at her job.

I met Sok when I visited STARTTS, the New South Wales Service for the Treatment and Rehabilitation of Torture and Trauma Survivors, in Sydney's west. For over 30 years, STARTTS has provided culturally sensitive psychological treatment and support to help refugees struggling with complex **trauma*** to rebuild their lives in Australia. Most of those who come to STARTTS have been exposed to multiple traumatic events, including war and violence, deprivation, and the loss of loved ones. Many have been subjected to torture or severe human rights violations.

Sok had come to STARRTS for therapy and **neurofeedback*** treatment. She had been having neurofeedback sessions for several months when I saw her, and she was feeling the benefits. She was more assertive, less fearful, and more aware of her tendency to overreact.

Mirjana Askovic, herself a former refugee, is a senior psychologist and coordinator of the neurofeedback program at STARTTS and she describes neurofeedback as a type of brain training.

'Neurofeedback uses technology to improve the brain's ability to self-regulate and then become more resilient. Sensors placed on the client's head measure the electrical currents from their brain, and real-time feedback on how their brain is functioning is provided in audio-visual form. In this way, clients can learn to recognise and control their brain activity and keep it within a specific set of electrical parameters linked to calm, attentive and relaxed states. No drugs or electrical currents enter the brain; the patient is simply learning to change the patterns of electrical firing to achieve better performance. Our brains are plastic and constantly evolving and, with enough training, the brain can learn these new patterns of functioning and make these changes permanent.

'Through the use of a quantitative electroencephalogram, also called "brain mapping", we observed that the brains of trauma survivors often have the reverse patterns of functioning. What does that mean? When these people need to be awake and alert to function in everyday life, their brains often go into 'slow mode', making them sleepy, tired, disconnected or dissociated. When they need to rest at night, their brains go into overdrive to keep them hypervigilant, making them agitated, anxious and angry.

These patterns were formed in response to traumatic experiences to help people survive and cope with painful emotions. To help clients to reverse these brainwave activity patterns that negatively impact their everyday life, we are using neurofeedback to alter these brain oscillations that are no longer functional.'

Askovic says that after about 20 sessions a client can notice some reduction in anxiety and tension and improvement in sleep. But further improvement takes time and practice. You need to have regular sessions, in the same way as you need to go to the gym regularly to build fitness. It varies from person to person, but Askovic says that about 40 sessions may be required before clients see significant change. There are no side-effects, and the changes appear to be permanent.

Neurofeedback and young people

An associate professor at Western Sydney University, **Dr Roger Gurr** is investigating how neurofeedback could be used to re-regulate the brains of traumatised young people. In late 2020 he invited Mirjana Askovic from STARTTS to join his program at Penrith headspace looking into this.

Multiple studies have shown that up to two-thirds of young people have been exposed to at least one traumatic event by the time they reach the age of 16. Those from refugee backgrounds and Aboriginal and Torres Strait Islanders are likely to have experienced much higher levels of trauma.

The period between the ages of 12 and 25 is a very important time for brain development. At puberty, the brain stops growing and begins to prune or remove connections that are no longer needed. This is when peer interaction and the drive to attract the best mate become really important for young people, so trauma at this time is particularly damaging.

Gurr says that during this stage of life, the brain is also highly plastic, so treatments like neurofeedback that encourage cognitive learning have an excellent chance of leading to improvement in this age group.

The neurofeedback project at Penrith headspace finished in October 2021, when the funding ran out. Gurr says that although the study hasn't been written up for a peer-reviewed journal and he'd like to have offered more sessions of neurofeedback, all the young people got some benefit.

Psychedelic drugs

The term 'psychedelic' refers to drugs that produce hallucinations, distortions of perception and apparent expansion of consciousness.

Psychedelic drugs came to prominence in the United States in the 1960s, when psychologist Timothy Leary developed the catchcry 'Turn on, tune in, drop out.' He was convinced that the psychedelic experience could be mentally therapeutic and even provide insights into consciousness itself.

But by 1966 psychedelics were demonised and banned. Now, in controlled scientific settings, we're seeing a new era of psychedelic research, exploring the potential of a range of drugs in mental health therapy, and how changes in consciousness created by these drugs can help us understand what's going on in the brain in different states of consciousness. But it's important to remember that most psychedelic drugs are still illegal today, and using them away from controlled medical settings, medically or recreationally, can be risky, both physically and psychologically.

In 2017 I attended a symposium in country Victoria that brought together researchers and others interested in the therapeutic benefits of psychedelics. I heard a range of stories about people's wild and dreamlike experiences taking psychedelics recreationally – but prominent, well-respected scientists from Australia and across the world were there to discuss the promising potential of psychedelic drugs to treat some types of mental illness.

Some scientists, like **Dr Ben Sessa**, psychiatrist and Head of Psychedelic Medicine at Awakn Life Sciences, say we are in the middle of a psychedelic renaissance, a term he popularised in 2012 with the publication of his book of the same name. He told me:

'When I was a youngster in the '70s, the drug education message was if you smoke cannabis, it's the same as injecting heroin. We are much more savvy than that now. We know what the killers are, we can see where methamphetamine and alcohol particularly are destructive to our society, crack cocaine, heroin. So we are no longer buying this government line of all drugs should be put in the same basket.' Secondly, he believes there is a general sense of malaise with traditional psychiatry, because many of the people who are prescribed psychiatric drugs find that they are not effective, or the side-effects are not tolerable. (Matthew, whom I introduced earlier, is one such person.)

'We've had this huge hiatus and now we're revisiting those studies of the '60s using modern neuroimaging techniques, and it's opening up whole new doors in terms of our understanding of consciousness ... The psychedelics represent the newest most innovative form of psychopharmacology for around 75 years. It's totally new. Unlike all of these other maintenance drugs, this is the first time that we are using these drugs in a sophisticated way to enhance psychotherapy, to get to the root cause of **trauma**.'*

Around the same time I spoke to **David Nutt**, Professor of Neuropsychopharmacology at Imperial College London, while he was in Australia giving a speech at the launch of Mind Medicine Australia, a philanthropic organisation set up to further psychedelic research here. He specialises in the research of drugs that affect the brain, and conditions such as addiction, **anxiety*** and sleep, about which he has written numerous papers and books. He is also Head of the Centre for Psychedelic Research at Imperial College, and told me he believes that psychedelic drug research has huge potential for treating mental illness.

He says the brain is the most complex organ in the universe and brain disorders are the largest burden of disease worldwide. The treatments we have are only moderately effective, and we're using treatments today that conceptually are around 50 years old, so it's time to try something different.

The main areas in which psychedelic scientific research is

currently focused are the use of **psilocybin,*** the active ingredient in 'magic mushrooms', and the synthetic psychedelic **LSD (lysergic acid diethylamide),*** in the treatment of **depression;*** and **MDMA (3,4-methylenedioxymethamphetamine),*** commonly known as the party drug **ecstasy**, for the treatment of **PTSD.***

LSD and psilocybin

'My first LSD experience just brought me deep into my feelings in a way that I had not anticipated,' says **Rick Doblin**, founder and Executive Director of the Multidisciplinary Association for Psychedelic Studies (MAPS), a not-for-profit US pharmaceutical company focused on developing marijuana and psychedelics into approved legal prescription medicine. 'And it wasn't always easy, the feelings were difficult, it was scary to be losing control of the ego, of who I was, of letting go and letting feelings well up in me. And I resisted it a lot.

'I wouldn't say that my first LSD experience was this profound mystical experience of connection, but it was the struggle between rational control and the emergence of these very powerful emotions. And it made me realise that this was a way to get in touch with my feelings and that the LSD was helping build my capacity for emotions to flow through me. I feel like LSD has fundamentally helped to balance me and also to inspire me.'

Professor Nutt shared with me some of the history of LSD. It was first synthesised in 1938, by the Swiss chemist Albert Hofmann, while he was researching a new blood stimulant for Sandoz pharmaceuticals. In 1943, with a hunch that the drug had something valuable to offer, Hofmann took LSD himself, and went on a famously 'magical' bicycle ride. His reports of his hallucinatory experience kickstarted a wider interest in psychedelics.

In the 1950s psychedelic research blossomed. It was realised that these drugs could be a very important tool for understanding and treating the brain, and about a thousand psychiatrists around the world started using LSD in research studies. There were many

good outcomes, with few adverse side-effects. 'Everyone thought
this was the future: we were going to revolutionise the treatment
of disorders like addiction and depression using LSD, and we
would have done if the drugs hadn't been made illegal.'

Over 15 years ago, Professor Nutt decided to resume the
research that ended when these drugs were banned in the late
1960s. He obtained permission to use psychedelics in a scientific
setting to explore what they do in the brain. He collaborated with
Amanda Feilding, the Countess of Wemyss and March, founder
of the Beckley Foundation, a charitable trust in the United
Kingdom that supports psychedelic research. In 2016 Nutt's team
revealed the first ever brain-imaging study of people under the
influence of LSD.

They were surprised to discover that these drugs do not turn the
brain *on*, they turn *off* the brain parts that control brain function,
known as the **default mode network*** (discussed in Chapter 4).
'If you see the brain as being a very complex orchestra, there are
several areas of the brain, particularly the frontal region and the
posterior region, which conduct the orchestra. And those are turned
off by psychedelics. And that then allows the brain to do things
that it hasn't done before, in the same way as if you take away the
conductor, they can stop playing Mozart and start playing jazz if
they want to, they can syncopate. And in a way under psychedelics
the brain can then syncopate and do things which actually it has
not been allowed to do ever since you were a child.'

In the brain scans Nutt's team conducted, one of the areas in
the frontal region that was switched off was the same area that
must be switched off to get people over depression. They thought
that if psychedelics were dampening this area down, perhaps they
could be used to help people with depression.

'So we applied for a grant, and to my amazement we got it,
the only money we've ever gotten from government to do this
research, but I think the fact we got it was testimony to the fact
that depression is the number one burden of disease in Britain, as
well as in Australia.'

A few months after my chat with Professor Nutt, I spoke to **Dr David Erritzoe**, who's part of Professor Nutt's Centre for Psychedelic Research. In trials conducted in 2017, the research centre team gave psilocybin to 20 patients whose depression had not responded to conventional treatment. 'And it was an oral dose of the drug we used, and therefore the duration is relatively long, so around four or five hours, so it's a long session with music playing and us being very present in the room sitting next to the person.'

The hope was that the participants would get some sort of new personal insight that would help them with their depression, and the results were quite positive.

'We saw that people's depression scores significantly dropped, in particular after one week,' says Erritzoe. 'I think 9 out of the 20, they were in remission after five weeks, and six of these had not relapsed, even after six months of follow-up. So in that sense quite impressive results.'

Since then the Imperial College team and other researchers around the world have continued to build on this research, investigating the potential of both LSD and psilocybin to make a real difference to those with depression.

Erritzoe concludes that it's important to do further research into psychedelic drugs like psilocybin for treating mental illness because the pharmaceutical industry has not produced many new compounds in psychiatry, and particularly not for depression, over the last decades, so it makes sense to look at all the promising alternatives. It's crucial, he adds, that the research continues to be scientifically rigorous, to show that psychedelic drugs are safe and effective for these mental health conditions.

MDMA and trauma

'I remember we were lying there in someone's house after going out and everyone was high, and we were in that youthful explorative, experimental-type way of thinking; we were all intellectual and we read Jung and all this sort of stuff.' Ben Sessa is describing taking ecstasy (MDMA) with his friends around the age of 18. 'And I remember we were lying there and we were saying, "Oh wow, this is so amazing! Try and think of the scariest thing you can, try and think of the worst thing in the world." And we all went, "Yeah, OK, let's try that, let's imagine our mums dying. Oh wow, it's not that bad, I could deal with that."

'And what's really interesting – I didn't know at the time; it was only in more recent years that I've come to find MDMA therapy – is that what I was doing there with my friends was essentially carrying out a little brief experiment of what we do with MDMA therapy. The drug provides this ability to engage with painful thoughts that normally you would avoid, but under the drug you can go there and do it and it's not that bad.'

MDMA is not a classical psychedelic like LSD or psilocybin, but it's been very popular in the nightclub scene. While it can produce feelings of increased energy, pleasure, emotional warmth and distorted perception, it can also cause unpleasant and sometimes dangerous effects. However, MDMA is being researched in controlled clinical settings as an adjunct to treatment for **post-traumatic stress disorder.***

As a psychiatrist who treats everyone from children to adults, Ben Sessa is interested in the trajectory from child maltreatment to adult mental health disorders like PTSD. In many cases, pharmacological drugs are not effective for treating trauma disorders, because they don't address the underlying pain.

What psychedelics offer, he says, is the opportunity to do focused and effective psychotherapy that gets to the root causes. People can begin to resolve these issues that have enslaved them, sometimes for decades.

Rick Doblin and his team at MAPS have also been researching the use of MDMA to treat severe PTSD. In 2021 they published the results of a large-scale trial showing that the treatment is relatively safe, and often extremely effective.

The therapeutic use of MDMA, he says, is 'based on a belief that there is an upwelling from the psyche that has healing potential and that we see that in dreams, that we all have this emergence into consciousness of repressed fears or anxieties or even just positive memories, that there is an inner healing mechanism'. The procedure involves three MDMA sessions over a period of eight hours, guided by two therapists.

'The classic Freudian concept is that the therapists let you free-associate and then they help interpret for you what's going on, and so *our* sense is that the expert is *you*, the person is *their own* therapist; *our* job is to support what is emerging, and because it's an eight-hour session we have the time and the patience for people to find their own ways.'

The effect of the MDMA is to reduce the activity in the **amygdala**,* where we process fear. The MDMA also connects the amygdala with the **hippocampus**,* where long-term memories are stored. Memory for fine details of what happened to the patient is enhanced, while previous fearful emotions are reduced, so that in a supportive therapeutic environment they can better face and accept the trauma as something that happened to them in the past.

MAPS acknowledge that their method could be improved, so they are funding studies with leading experts in other forms of therapy, and helping them blend MDMA with their existing methods.

Why is Doblin so passionate about this research? He told me: 'I think the world is in danger … We need to see our interconnectedness, and once we see our interconnectedness we are not so easily willing to demonise others that are different from us. And I think that that's the long-term strategy that I'm working on, and I feel that there are all different ways that people can have those kinds of experiences …

'And so I think the reason for doing the medicine and the research is to try to open the culture up so that eventually there can be these opportunities for more and more people. I think we need broad-based mental health improvements, and I think psychedelics can contribute to that.'

20

REVOLUTIONARY NEW TREATMENTS FOR DEMENTIA

Conversations with:

Lee-Fay Low, Professor in Ageing and Health, University of Sydney

Joanna Weinberg, musical writer, performer, choir director, actor, songwriter

Muireann Irish, Professor of Cognitive Neuroscience, School of Psychology and Brain and Mind Centre, University of Sydney

Laughter in dementia care

'Twenty years ago dementia was a hidden issue,' says Sydney University professor **Lee-Fay Low**. 'And then we had a few medical breakthroughs and there were a few drugs for dementia. And then it became reasonable to diagnose dementia because you could give them a drug treatment for it, even though the drug treatments aren't very effective, and they probably came to Australia in the late 90s.

'Since then we have been promised more medical breakthroughs and none of them have come. So I think the shift has been to look for *non-drug* ways of looking after people with dementia, and the way you care for them then becomes even more important. Every breakthrough that's promised isn't going to come, and they say 5 years, 10 years, 20 years, and we are still waiting for that. So the way we care for people with dementia is critical, then.'

With an ageing population, and with dementia being the single greatest cause of disability in people over 65, good and respectful dementia care is a very real challenge for our society. And Lee-Fay Low and others are working hard to meet this challenge.

In 2013, I was lucky enough to accompany actor and entertainer **Joanna Weinberg** on her mission to make some fun in the secure dementia unit of a Sydney aged-care residence. Many of the residents' faces lit up with smiles as they saw her come in with her maracas, xylophone, shakers and bubble maker. Joanna was wearing a wonderful maroon-coloured jacket trimmed with gold braid and brass buttons, with a matching maroon fez on her head.

Joanna was one of the clown performers from the non-profit Arts Health Institute's Play Up program, using humour and fun to improve the lives of people with dementia. Play Up grew out of a collaboration between Jean Paul Bell and other Clown Doctors at the Humour Foundation of New South Wales, Professor Lee-Fay Low, and the Sydney-based Dementia Centre for Research Collaboration.

Professor Low told me that while antipsychotics are used for aggression and agitation in people with dementia, some of the side-effects can be unwanted motor movements, and increased risk of falls, stroke and even death. Also, research now suggests that antidepressants don't work for **depression*** in people with dementia, so all the guidelines recommend trying non-pharmacological treatment first.

She points out that 'an engagement or an activity-focused approach really gives them reasons to keep doing things, and there's more and more evidence that suggests that if you keep people with dementia active – mentally active, physically active, socially active – that can improve their mood and possibly delay the progression of their symptoms'.

Professor Low and her colleagues became inspired by anecdotal evidence from the use of humour therapy in nursing homes in the United Kingdom. They collaborated with the Humour Foundation and conducted the Smile Study over a three-year period. It involved about 400 residents from 36 nursing homes, randomly allocated to two groups. One group was given the humour therapy intervention, and the other continued their usual care. The intervention involved 12 weekly visits by professional clown performers. The residents were observed to be happier after the sessions and behaved more positively towards each other as well.

The Smile Study showed that there were no significant differences in levels of depression between the humour therapy group and the control group, which could have been because the researchers didn't specifically select residents with depression. But the study also found that the level of agitation significantly decreased in the therapy group.

Professor Low told me, 'We think that behavioural and psychological symptoms of dementia such as agitation happen because people with dementia have unmet needs and usually they can't express what these needs are. So they may be physical needs, such as needing to urinate or being hungry or cold, and often their needs are for company or stimulation, because if you imagine what it's like living in a nursing home, it would be really boring. People look after you physically, but nothing else much happens and you just kind of sit around, and the television might be on, if you can hear it, and it's all pretty dull. So we think that we actually met an unmet need for company and interaction, and that's why the levels of agitation decreased.'

I was keen to hear more about how Weinberg connects with the residents. 'I start from the point of empathy and compassion,' she told me. 'I try and embody joy in my being. Then I move into all my skills, so my skill set is music, singing, I've got a very strong theatre background, comedy, and I'll try and elicit a response of laughter, connection, anything to try and help the person who I'm interacting with feel that they are human and that they are worthwhile. It seems universally that everybody that I work with enjoys music … They may not remember where they are today, why they are here, who their family are, they may not remember the names of their children or their friends, but they can sing "Daisy, Daisy". It's quite extraordinary.' (In Chapter 22, 'Music in Mind', I'll share with you some moving examples of how powerful music can be for people with dementia.)

Weinberg also uses storytelling to connect with people, and when we were in the dementia care unit, she introduced me to a woman we'll call Mary.

When Weinberg first met Mary, she was very introverted, always slumped down in her chair and often asleep. Weinberg found from her file that she used to be an English teacher, so she started to read poetry to her.

One day Mary started reciting a poem back to Weinberg from memory. It turned out that as a young woman, her speciality was the poetry of Keats, Byron, Coleridge and Shelley, and hearing the poems again triggered her memory. Then she started reciting whole swathes of poetry by Keats to Weinberg. They became friends over their shared enjoyment of English literature, and Mary has changed profoundly since Weinberg met her.

When I chatted to Mary, she quoted a line from Shelley's 'Ode to the West Wind', then apologised because she couldn't remember the rest. She went on to tell me that she had once met Welsh poet Dylan Thomas, with his wife and another woman. She said the memory about him she treasured most was that he had a beautiful reading voice.

'Anything you do with someone with dementia has to be tailored to that person,' Professor Low points out. 'So it has to be based on their life, their personality and what they want. I think you have to give them choice and control. So if they choose to not want to engage with our humour therapists, say, that's OK, you've just got to try again next time. I think we should give them back status. You lose a lot of status as you grow older, and if you've got dementia and you are in a nursing home you lose even more status. So just giving them an opportunity to give back as part of the activity, I think that's really important as well.

'And I think that it should be engaging and it should be fun because, if not, there is no reason for them to do it. They don't have a very good ability to plan for the future, if that makes sense. They live in the moment because their short-term memory is so bad. So if you say, "You should exercise because it's going to be good for your health", that doesn't make any sense. But if you say, "You should dance because it makes you happy now", *that* makes sense.'

Low describes a series of fun activities called cognitive stimulation therapy which has been run for people with mild to moderate dementia. It's all about doing fun activities – playing games, lawn bowls, singing songs, doing crosswords together – for at least 90 minutes a week in a group setting. This therapy has been shown across many studies to improve or at least maintain cognition compared with not getting that type of care.

There are a few studies showing that getting people to practise doing functional or self-help activities, like preparing a sandwich, making their bed and folding sheets, can also help to maintain cognitive function. And smaller studies are looking at activities like pet therapy and doll therapy.

Low comments: 'It's all looking very promising towards the picture of the person having a meaningful lifestyle while living with dementia.'

The Clown Doctors continue to run a program; the Arts Health Unit is no longer operating, but some of the artists who were trained with them continue to work under the banner of the Outside In Collective.

Funding for the Dementia Collaborative Research Centre ran out, but many of the researchers continue to work in the dementia space. Lee-Fay Low currently chairs the Sydney Dementia Network, of which Professor Muireann Irish (see below) is also a part.

Dementia, daydreaming and the future

As Low observed, people with dementia have a poor ability to plan for the future. And this question of how dementia patients conceive of the future has also intrigued **Professor Muireann Irish**, whom I introduced in Chapter 6 in connection with daydreaming. When we met, Irish and I both shared our own experiences with dementia: I spoke about my mother's and father's dementia, and Irish told me that watching her grandmother live with Alzheimer's disease prompted her to specialise in memory and cognitive changes in dementia.

In one area of her research, Irish has revealed more about the inner experience of those with a particular form of dementia called frontotemporal dementia. 'Frontotemporal dementia is a younger-onset dementia where the cardinal complaints are often a change in personality and in behaviour. So these people might start to act out in seemingly inappropriate ways; they may appear disinhibited, impulsive and not able to regulate their behaviour in different social contexts. And so it can be very distressing for family members to manage these changes in behaviour, particularly as the person living with dementia often may not have insight or be aware of their actions.'

It was often assumed that people living with dementia were lost in their own world of mind-wandering, or daydreaming. But no studies had formally tested this, so Muireann Irish decided to investigate.

It was a challenging task for Irish and her team to measure something as complex and introspective as daydreaming in people with frontotemporal dementia. First, they created a task that was monotonous and boring – highly conducive to mind-wandering. Participants were told to relax and to look at a series of simple shapes that appeared on a screen for varying durations. Then the participants were asked to report on what they were thinking about during that time.

The results were absolutely remarkable, according to Irish. There was a complete lack of rich, vivid mind-wandering. The participants with frontotemporal dementia either reported that their minds were blank or they simply commented on the perceptual features of the shapes; for example, that they saw a red square, or that they liked the colour red. Their thoughts were very much tied to what was immediately in front of them in the environment. They seemed to lack an ability to disengage from the stimulus and spontaneously generate their own internal monologue.

'There is something unique about the way we toggle between our inner thoughts and the external environment, and this capacity requires highly sophisticated brain networks to work together in concert. What we now know in frontotemporal dementia is that key nodes of these brain networks are compromised, disrupting the ability to seamlessly switch between the external and internal worlds.

'Essentially, patients lose their ability to daydream, which has several knock-on effects. They show difficulties thinking about other people's perspectives, and seem particularly impaired at looking forward to events that might occur in the future. This can deeply affect their self-identity and lead to apathy and feelings of depression.'

This insight has important implications for the way we care for people with dementia. It means, Irish suggests, that we have to make every effort to provide a stimulating external environment for people living with frontotemporal dementia. 'We can't assume that if someone is left to their own devices, they will have the capacity to spontaneously entertain themselves via their own internal world,' says Irish.

Irish's study also suggests that people with other forms of dementia, such as Alzheimer's disease, don't necessarily lack this ability for spontaneous thought – so what may be compromised in one group of dementia patients may be relatively intact in another.

As Low also observes, this all points to the need for highly personalised care in dementia institutions. A one-size-fits-all approach is not appropriate. Every person with dementia has their own history, ideas and outlook on life. So care needs to be carefully tailored in order to promote real wellbeing in each individual.

21

THE MIND–BODY CONNECTION

Conversations with:

Jo Marchant, freelance science journalist and author

Shannon Harvey, journalist and filmmaker

Professor Craig Hassed, Senior Lecturer, Monash University, and Director of Education, Monash Centre for Consciousness and Contemplative Studies

Honorary Professor George Jelinek, founder of the Neuroepidemiology Unit, School of Population and Global Health, University of Melbourne

Suzanne O'Sullivan, neurologist, National Hospital for Neurology and Neurosurgery, London

Lorimer Moseley, Chair in Physiotherapy, Professor of Clinical Neurosciences and Bradley Distinguished Professor, University of South Australia

Associate Professor Sylvia Gustin, School of Psychology, University of New South Wales

Tasha Stanton, Associate Professor in Clinical Pain Neuroscience, University of South Australia

The power of the placebo

From ancient practitioners of philosophy and religion to modern scientists, humans have been contemplating the **mind–body connection (MBC)*** for thousands of years. Yet the idea that our minds can heal our bodies has always been a controversial one. Mainstream contemporary science and healthcare have tended to treat the mind and body as separate entities and have associated the mind–body connection with new-age spirituality and the realm of health gurus. However, scientific research is now revealing that there is a bidirectional relationship between the body and the mind, and that our thoughts, emotions and subjective experience can indeed have beneficial impacts on our physical health.

Jo Marchant, author of the bestselling book *Cure: A Journey into the Science of Mind over Body*, decided to put aside her scepticism and explore the scientific evidence now coming to light for how our mental state can affect our health. Marchant told me she began by exploring the **placebo effect*** because it's a very pure example of how the mind can affect the body. It occurs when a person's physical or mental health seems to improve after taking a treatment they believe to be effective, even though it is a fake, or a placebo. It's long been assumed that this effect is an illusion, but scientists are discovering that a placebo can trigger measurable biological changes in the brain and the body that are similar to those caused by drugs.

For example, Marchant says, when someone takes a fake painkiller, there is a release of endorphins in the brain. These are natural pain-relieving chemicals that opioid drugs like morphine and heroin are designed to mimic. If your pain is eased after taking a fake pill, the same biological mechanism is being activated as when you take a drug. It's not in your imagination, it's really happening.

People with Parkinson's disease, for instance, lack the neurotransmitter (chemical messenger) **dopamine**. When patients with this condition are given a placebo, a flood of dopamine is released into their brain that relieves their symptoms. Likewise, giving people with altitude sickness fake oxygen reduces their levels of the chemicals behind many of the symptoms of altitude sickness. There are many different mechanisms involved in the placebo effect that act using the same biological pathways as drugs would.

Marchant says there is some understanding of why the placebo effect occurs. It is related to the fact that symptoms like pain, nausea and fatigue are warning signals telling us that we need to change our behaviour.

Marchant has also investigated how the power of the placebo can be harnessed in routine medical treatment. One of the problems is that it's not considered ethical for doctors to lie to patients, telling them they are getting effective drugs when they are not. But there is research showing that 'honest' placebos – placebos the person knows they are taking – are still effective. This has been shown in a few studies of conditions such as **depression,** * irritable bowel syndrome, headaches and ADHD. Honest placebos are not as effective as real placebos, but they are significantly better than no treatment.

To incorporate the power of the placebo effect into conventional medicine, it's important to understand what triggers it. Studies suggest that expectation is very important. What you are told about the treatment, how you will benefit, and what the side-effects might be can make a difference.

Also important is the doctor–patient interaction. Marchant found studies showing that patients with a wide range of conditions do better if the practitioner is warm and empathetic rather than cold and removed, regardless of what treatment they are getting. Longer, more interactive consultations are also more effective than shorter, more standardised ones.

In her research, Jo Marchant met Ted Kaptchuk, Professor of Medicine and Professor of Global Health and Social Medicine at Harvard Medical School. He began as a practitioner of traditional Chinese medicine, including acupuncture, but he began to realise that his patients were showing dramatic improvements before he'd given them any treatment at all. To find out more about the mechanisms of the placebo effect, he conducted trials in which everybody was given placebo acupuncture – the needles were put in the wrong places and didn't fully penetrate the skin. He investigated how the attitude of the practitioner affects patient recovery and compared placebo acupuncture with a placebo pill.

'And you might think,' Jo says, '"Well, there shouldn't be any difference, because there is no active ingredient in either of them", but actually it makes a *big* difference. When he tested those on people with chronic arm pain, he found that acupuncture was better for the pain but the placebo pill was better for helping people to sleep.

'So he is really coming up with this new way of looking at placebos, where it's not about the placebo itself, that the key to placebo responses is really in the patient's brain, it's our psychological response to receiving that treatment, what that treatment means to us.

'And that could also explain things like why placebo effects vary in size between different countries, for example; there are real cultural effects. Or why different coloured pills can have different effects. So, for example, blue pills tend to have a more calming effect, whereas red pills are better for pain – except in Italy, where blue pills among Italian men are not calming at all, they tend to be more of a stimulant, and researchers think that might be because blue is the colour of the national football team in Italy.'

But placebos can have detrimental effects too. 'There are lots of examples where believing that something is going to harm us can actually create those very symptoms,' Jo notes. 'So it's often mentioned in association with things like voodoo curses, you know, the stories of people falling terribly ill and even dying if they thought that they had been cursed.' This is called a **nocebo**.

This can also affect us in everyday life. Sometimes the side-effects that we experience when we take drugs aren't actually due to the drugs themselves, they are due to our expectation of experiencing those symptoms.

Conditioning the brain to heal the body

In the early 1980s, Marette, an 11-year-old girl from Minnesota, was diagnosed with lupus, a chronic autoimmune condition. It caused her immune system to attack almost every tissue in her body. She was treated with steroids but became increasingly ill over the next couple of years, until her heart started to fail. Her doctors wanted to give her a toxic immunosuppressant drug, but her mother was worried that this drug could be as dangerous as the condition itself.

Jo Marchant told me that Marette's mother heard about conditioning research that had been done by Robert Ader at the University of Rochester, using rats. In his research, he gave the animals sweetened water, which the rats love, at the same time as injecting them with a toxic immunosuppressant drug, which made them feel sick. Later when the rats were given sweetened water only, they refused to drink it. He then force-fed them the harmless sweet water using an eyedropper and the rats did not forget their aversion to it, so much so that one by one they died. Ader concluded that the harmless sweetened water suppressed the rats' immune systems so extremely that they became vulnerable to fatal infections. His results showed that not only do learnt associations affect responses like nausea or heart rate, known to be regulated by the brain, but they can also influence immune responses.

This is the line of research into conditioning, which investigates how external cues can prompt physiological responses – as with neutral psychological cues like sound, taste and smell influencing the immune system. The research uses the idea of learnt associations, borrowed from the famous Pavlov's dogs experiments, where dogs were trained to associate a sound or a light with being fed, until just the sound or light on its own was enough to make them salivate.

Marette's mother arranged for Ader's team of researchers to collaborate with Marette's doctors to design a special conditioning regime for her, in the hope that they could train her immune system to respond to a lower drug dose than normal, thereby minimising her exposure to toxicity. Once a month for three months, she was given the immunosuppressant drug through a vein in her foot while she sipped cod liver oil and breathed the scent of a rose perfume that was sprayed into the air around her. After that, she was exposed to the cod liver oil and rose perfume every month but only received the toxic drug every third month. By the end of the year, the doctors reported that Marette had responded as well as she would have after the full drug dose regime. As a result, she was able to get through her health crisis with a much lower dose of the drug.

As Marchant explains, 'There have been several trials now in animals and in human volunteers showing that you can use these processes to reduce drug doses with the same effects. So there are studies going on in kidney transplant patients now. For example, I met one kidney transplant patient who was taking part in this trial and they were drinking a mix of strawberry milk, green food colouring and lavender oil alongside their immunosuppressant drugs, and then seeing if that drink on its own could suppress the immune system.

'The idea that these neutral psychological cues – sound, taste, smell – can actually influence the immune system, that was a real surprise for me.'

She adds: 'Researchers have said that for everyday conditions, that is something that might help to reduce drug doses. I think for the more serious life-threatening conditions we are going to need a lot more research on that, but they are very hopeful that in a few years' time this is something that we might be able to use for everything from autoimmune disease to organ transplant patients to cancer, to reduce the side-effects and toxicity of those drugs with the same clinical benefit.'

Shannon Harvey and *The Connection*

Journalist and filmmaker **Shannon Harvey** was diagnosed with an autoimmune disease when she was in her early 20s. She was told by doctors that if the disease progressed, she could end up in a wheelchair or with organ failure before she turned 30. After trying all manner of conventional and alternative medicines, to no avail, she decided to investigate the mind's capacity to heal the body.

The stress and relaxation responses

One of the first people Shannon turned to was **Professor Craig Hassed**, Director of Education at the Monash Centre for Consciousness and Contemplative Studies. His interest in mind–body medicine began when he learnt about the placebo effect as a student.

It's becoming more widely accepted that stress can significantly influence our physical health. Hassed explains the mechanism at work. Stress, and other difficult mental and emotional states such as anger, hostility and depression, are all associated with activation of the **sympathetic nervous system**, which mediates the **fight or flight response**. (I looked at this response back in Chapter 12, in connection with anxiety.)

Stress, of course, also has a useful role to play in many aspects of our lives. As an example, Hassed describes a scenario in which you are taking your morning walk through the park and you come across a tiger, escaped from the zoo and looking for breakfast.

This immediately activates the fight or flight response. Your heart rate and blood pressure increase, causing extra blood flow to your muscles to help you get out of danger. Sugars and fats pump into your bloodstream and your metabolic rate goes up, so you feel hot and begin to sweat to keep yourself cool while you are exerting yourself. You go pale as the blood is diverted away from your skin and gut, so your gastrointestinal system shuts down. Your blood thickens and will clot faster than normal, which could mean the difference between life and death if the tiger gets a hold of you. Your immune system is pumping out inflammatory chemicals. You are very mindful, and your brain is relatively quiet but very focused.

This stress response is designed to save your life, and is vital on the rare occasions when it's needed. 'But when we activate that response all the time when we don't actually need it,' adds Hassed, 'by anticipating future events or replaying past events, getting overly angry and reactive to events even if they are happening in day-to-day life, what we do is we *overactivate* this fight or flight response, and it has a long-term cumulative effect that's called **allostatic load**.' This accelerates ageing and associated illnesses, right down to the DNA in our cells.

The opposite of the fight or flight response is the **relaxation response**, which allows us to come back to rest once we no longer need either to fight off or to flee from a tiger or other source of danger. It's your personal ability to encourage your body to release chemicals and brain signals to make your muscles and organs slow down and increase blood flow to your brain.

'Relaxation response' is a term coined by the late Dr Herbert Benson, former Mind Body Medicine Professor of Medicine at Harvard Medical School and Director Emeritus of the Benson-Henry Institute for Mind Body Medicine at Massachusetts General Hospital. Shannon Harvey told me Dr Benson was one of the most remarkable scientists she spoke to when researching MBC.

The power of meditation

The relaxation response came out of research Benson did in the late 1970s and early 1980s, in which he noticed consistent changes in the brain as a result of **meditation**. Cutting-edge science has continued from his early work to confirm that meditation elicits the relaxation response, including by turning down the genes that affect disease. (This is part of the field of **epigenetics**,* which I looked at in Chapter 17 when discussing **intergenerational trauma**.*) This idea excited Harvey, who had been led to believe that her autoimmune disease meant she was destined for a lifetime of poor health.

Hassed too is a strong proponent of meditation and **mindfulness**,* and feels indebted to Benson's work. 'In a lot of ways he was very prophetic,' Hassed says, 'because the research has gone a lot further than those initial early studies, but it really just documents the importance of that relaxation response – that we need to learn how to recognise the inappropriate activation of the stress response and switch it off.'

Hassed explained to me the effect that meditation has on our brains. When we are distracted or apprehensive, the brain slips into **default mode**, and executive functioning areas involved in information processing, decision-making and emotional regulation do not work well. (I looked at the **default mode network*** in connection with creativity, in Chapter 4.)

When default mode activity is high, the brain produces a lot of **amyloid**, the protein that's thought to contribute to **Alzheimer's disease**.* Hassed points out that research shows practising meditation counters this apprehensiveness and changes the brain: grey matter of the brain thickens and there is new neuronal growth, especially in the executive functioning and emotional regulation areas. The brain's stress centre, the **amygdala**,* quietens down quite significantly.

Recent research is suggesting, says Hassed, that mindfulness and meditation may be very important for preventing cognitive decline associated with ageing, and for maintaining a healthy brain. (I'll have more to say about the positive effects of meditation in Chapter 23.)

Reversing cellular ageing

American physician Dr Dean Ornish was another influential person Harvey met on her healing journey. Ornish is known for his Program for Reversing Heart Disease, which has proven effective in treating not just heart disease but early-stage prostate cancer as well.

'But the thing that really excited me about meeting him and interviewing him,' says Harvey, 'is that his research is actually looking at this idea of **telomeres**.' Ornish teamed up with Tasmanian Elizabeth Blackburn, who won the Nobel Prize for discovering them.

'Telomeres are related to cellular ageing, so the best way to describe what they are is that they are little caps that sit on the ends of our chromosomes like the ends of our shoelaces, and as we get older the ends of those shoelaces get smaller and smaller and smaller. And what Dr Blackburn and Dr Ornish found is that when they went through that Ornish program, which involved lifestyle changes, things like meditation, yoga, family support, community support, as well as diet, they discovered that people's telomeres actually got longer.'

MBC and multiple sclerosis

Harvey was also inspired by **Honorary Professor George Jelinek**, founder of the Neuroepidemiology Unit at the University of Melbourne's School of Population and Global Health. I spoke to Jelinek, who told me that he was devastated when he was diagnosed with multiple sclerosis (MS). The disease has a strong genetic component, and many years earlier Jelinek's mother had taken her own life because her MS symptoms had become unbearable. 'It really felt like an enormous hand just reached into my life and plucked my future away in front of my eyes. I could immediately see that descent into paralysis and catheters and walking frames and wheelchairs, and all the things I'd seen with my mother suddenly became my future. And I just felt that it was all gone.'

He was determined not to accept this diagnosis, for which conventional medicine has no cure, as destiny. After meeting with an old friend and mentor, 'I realised that I'd so compartmentalised my life around work and my obligations that I hadn't really given time to some of the more critical things that I needed to think about and needed to explore in my life.'

He used his academic background to search the medical literature for answers to his health condition, and was surprised to find that it is possible to remain well after a diagnosis of MS. 'So I found things like diet ... I looked at the benefits of sun exposure in autoimmune illness ... I exercise regularly ... I now have started meditating again regularly ...

'But there was also ... the spiritual side of my life, and I started exploring that in detail ... And I started reading a lot more about healing ... And I started meeting people who – I went out of my way to meet people who I knew had recovered from illness or who I found to be uplifting in a spiritual sense ... And that combination of things really has all interacted and contributed to keeping me well.'

Not only did he recover from MS, but he also shared his recommendations with others through the development of his Overcoming Multiple Sclerosis program.

Acceptance of MBC

I asked Hassed: is the holistic approach of MBC being incorporated into conventional medicine?

He told me, 'The uptake has been far too slow. One of the reasons is that it's not a patentable product, so you don't get the marketing push that you do for other interventions and treatments that are patentable and much more marketable.

'The other part I think has to do with education. Unfortunately I think most of our medical schools are decades behind the research in this area. So it doesn't get integrated into medical practice. I think the research agenda needs to catch up as well.

'There's an absolute goldmine down there if we could plumb the depths with just a few more resources in that area.'

We can only hope that this situation changes as the scientific evidence behind the mind–body connection becomes more widely understood.

The Connection

Inspired by the mind–body research she came across, Shannon Harvey incorporated this knowledge into her own life. She also produced a documentary about her findings, called *The Connection*.

Not long after it was released, a routine blood test showed no sign of autoimmune disease in her body. However, she is careful to describe her experience as 'recovery' rather than 'cure', as she still has the genetic predisposition for an autoimmune disease and has had a number of setbacks since her recovery.

'A lot of people ask me what's the magic bullet. You know: "I've been diagnosed with something, what's the one thing that I can do in order to get better?" And I'm really very firm in saying there is nothing simple about it, and it's also not easy, it's hard work.

'When I think back on my life previously when I was first diagnosed and I compare it to my life now, it's almost unrecognisable. I'm a completely different person.'

Imaginary illness

Brenda was experiencing a very prolonged seizure that was not responding to medication. It was assumed that she had epilepsy. She was taken to an intensive care unit, where she was ventilated and given strong sedative medication, but every time the sedatives were withdrawn, she began having further seizures. She was referred to a neurology consultant, who concluded that Brenda's seizures were **dissociative** – coming from her subconscious.

Suzanne O'Sullivan is a neurologist at the National Hospital for Neurology and Neurosurgery in London. She's developed a specialist interest in the mystery of **psychosomatic illness*** after seeing patients like Brenda.

'When I encountered that in the first instance,' O'Sullivan says, 'as a very junior doctor, not fully understanding these sorts of disorders, it was tempting for me to relate dissociative seizures to meaning there was nothing wrong with her. So it was hard for me in the first instance to appreciate that a seizure that comes out of the mind is every bit as disabling and life-destroying as a seizure that comes through epilepsy, and that's something I had to learn by meeting a lot of patients and seeing how often their lives are destroyed by this diagnosis.'

Brenda *did* make a recovery through psychological support, because the correct diagnosis was made quite early in her illness.

Another case that really moved O'Sullivan as a junior doctor was that of a woman who, on examination, was blind. 'When you examined her, she was densely blind, so she could see slight changes in light and make out some shapes, but really couldn't see anything more detailed than that,' O'Sullivan explains.

'The more I got to see her the more I felt like she was looking directly at me when we were in conversation. I began to notice she was doing things that looked like she was sighted, so she would reach out and pick up objects, for example, which I felt she couldn't do unless she could actually see them ...

'On the day she left the hospital she gave me a card that she had made on which she had drawn a picture, and the picture was perfect; it was extremely difficult to appreciate that it had been drawn by someone who experienced herself as being densely blind ...

'There was an extreme innocence in this woman's behaviour which really convinced me that she had no insight into the split between what she could see and what she was aware of seeing. I was also very influenced by the way she pursued a diagnosis. She was very keen to have investigations, as all of these patients are, because they believe that there is a physical disease causing their disability. And someone who is pretending to be ill or faking illness doesn't pursue a diagnosis in that way. So meeting people like her has really helped me to see how severely people suffer and how desperate they are for an explanation for what is happening to them.'

Psychosomatic illness vs hypochondria

Psychosomatic illness is *not* **hypochondria**. O'Sullivan explains the difference: 'Hypochondria is anxiety about illness. So people are disabled by their anxiety, as opposed to by the symptom they have. So the symptom itself might be very minor, it could be a very small headache or a very minor feeling of shortness of breath. The reason that the person can't function is because they are so anxious about the symptom.

However, 'a psychosomatic illness is a disorder in which people have real disability with significant level of suffering but where the disability cannot be explained by medical tests or by physical examination, the point being that the disability itself is real, despite the lack of explanation for it. Often there is no coexisting feelings of anxiety or psychological problem going along with it, the predominant feature of a psychosomatic illness is the disability rather than any psychological suffering.'

Features of psychosomatic illness

As a neurologist, O'Sullivan looks after people with diseases of the brain, nerves and spine, but a significant proportion of her patients have neurological symptoms such as seizures, headaches, numbness or tingling that have no medical explanation.

The commonest types of psychosomatic illness are pain and fatigue, but she sometimes encounters more unusual and extreme forms, such as seizures and paralysis. We can only really explain *some* of the symptoms people get. Some are related to the activation of the **autonomic nervous system**, which controls the **fight or flight** and **relaxation responses** I looked at earlier, or to other body processes. More extreme symptoms can't be explained by *any* bodily mechanisms.

However, O'Sullivan adds that we know the brains of people with psychosomatic illness are processing information differently from other people's brains. Brain scanning has shown that different areas of the brain are activated in people who have psychosomatic illness from those activated in people who are well, and people who are faking illness.

The disorder has a variety of causes. 'I think for a percentage of people there is **depression*** or **anxiety,*** and a percentage of people have suffered a significant **trauma,*** sometimes in the distant past. So sometimes these illnesses manifest themselves years after the trauma, and that occurs because people have buried that trauma deep inside and it takes a long time for it to make itself known. But that isn't the case for all people. For some people it is more of a behavioural disorder in which their symptoms arise as a result of the way they respond to injury or illness or stress at work or stresses in their home life and so forth.'

Of course, once a person is diagnosed with a psychosomatic illness it can be very difficult for them to accept that their illness is psychological. Explaining it to others is also challenging, as there is so much stigma around it. It's actually much more common than is often thought, because people generally don't talk about it.

'A GP would see multiple people every day. Approximately one-third of the patients they see every day have medically unexplained illness; the same applies to specialist clinics like neurologists, rheumatologists, gynaecologists ... So these sorts of symptoms in their less severe form are extremely common, and many of us will be affected at some point in our lives, but hopefully only in a minor and transient way.'

She adds that 'women are much more likely to be affected by psychosomatic illness. So two-thirds of the people who have seizures as a form of psychosomatic illness are women, and the same applies to most of the forms of the illness. It's very difficult to say exactly why that is the case. Some of it may be culturally determined. Men and women have a different way of expressing distress, and it may be that it's more culturally acceptable for a woman to express her distress in that way ...

'The other possible explanation is that women are more likely to suffer the sorts of abuse that can lead to psychosomatic illness, so things like physical abuse and sexual abuse are seen in higher rates in people with psychosomatic illness.'

Because the causes are so diverse, individualised treatment is required, says O'Sullivan. Some may respond to seeing a psychologist or psychiatrist, but in others the symptoms should be treated as physical illness.

O'Sullivan suggests that a shift in perspective is required to better understand and treat people with psychosomatic illness. 'I think that people consider medicine to be much more scientific than it is. Medicine is based on science, obviously, but the practice of medicine

I think is much more of an art ... Illness is the human response to disease. And everyone's response to disease and everyone's response to what happens to them is completely different, and that's really what makes diagnosing diseases and illnesses so difficult.'

She argues that psychosomatic illness needs to be resourced in the same way as less stigmatised forms of illness. 'What we really need are multidisciplinary teams. We need neurologists who are interested in this condition who work alongside psychiatrists and psychologists who are interested in this condition, and also work alongside people like physiotherapists, occupational therapists and social workers. In every other speciality, if you have arthritis or heart problems you have dedicated teams for each of these disorders, whereas psychosomatic disorders are very neglected, and I think we need more multidisciplinary care pathways.'

Above all, she says, 'I think really what we need to do is stop thinking of things like the brain and the mind as being completely separate.'

Pain on the brain

Scientists are now realising that the brain plays a crucial role in how pain is experienced, and it's opening the way for some innovative treatments.

Acute pain is a normal sensation that alerts us to injury and passes once the body is healed. But **chronic pain** is very different. It often persists when the cause is no longer evident.

It's thought that one in five Australians lives with chronic pain, and over the age of 65, it's more like one in three. In the search for more effective treatments, scientists are discovering more about the key role that the brain and the mind play in our experience of pain.

Lorimer Moseley is Chair in Physiotherapy and Professor of Clinical Neurosciences at the University of South Australia, and leads the Body in Mind Research Group, which is investigating the science of pain. He says that the most significant contribution to this field in modern history was made in 1965 by the late neuroscientist Patrick Wall and the late psychologist Ron Melzack, 'and apparently they conceived of this in a pub on a napkin, which is pretty cute'.

They proposed that 'there is a type of message gate in the spinal cord that can be opened and shut to let messages through from the tissues of the body. And these messages are not pain messages, they are danger messages. But the really important advance in that work was their statement that that gate in the spinal cord is controlled in part by the brain. So the brain sends messages down the spinal cord to open and close the danger gate, if you like ... The brain is ultimately the big kahuna in making the decision about what's biologically advantageous for you as an organism.'

Any evidence that your tissue is in danger will increase the likelihood and intensity of pain, and evidence that your tissue is *safe* will *decrease* the pain.

It's now understood that these pain cues are more important in *chronic* than *acute* pain. Moseley points to some interesting work demonstrating this, looking at the relationship between getting a scan or X-ray of your back. Your pain will tend to be worse after having the scans, because you interpret what's seen in the scans as abnormal and dangerous.

'If I can give you an example, we did an experiment 10 years ago where we got a very cold stimulus and put it on the back of the hand of a group of healthy volunteers, and we simultaneously showed them a light-blue light or a red light. And then we got them to report how much it hurt ... What we showed is that when they get the very cold stimulus with a red light, it hurts more. And even more remarkable, most people describe it as a hot stimulus, not a cold stimulus.

'So that red light has provided credible evidence that the tissue is in more danger because red means hot and hot is more dangerous ... And those sort of findings really have kicked off a revolution in our appreciation of how complex the biology of pain really is.'

Moseley explains chronic pain using a metaphor of the brain as an orchestra: 'If that orchestra just keeps playing one tune over and over again, it gets stuck on the tune. It becomes very efficient at playing that tune and it becomes less able to improvise, it becomes less efficient producing other tunes. And in that metaphor, the tune that is repeatedly played is pain. And the system becomes better at producing pain.'

This is a normal biological adaptation as our body tries to protect us, but it often means we are being protected from something that we don't *need* to be protected from. While analgesic drugs provide pain relief for many, they often have adverse side-effects and lead to tolerance issues and a risk of addiction. These are the challenges faced by pain specialists.

Associate Professor Sylvia Gustin from the University of New South Wales has been researching chronic pain for over 20 years. She described two of her patients to me.

One had a very painful tooth, so she went to the dentist, who removed the tooth and the wound was healed. However, the patient's oral facial pain continued.

Another sustained a spinal cord injury from a motorbike accident. Afterwards he woke up from a coma with very strong pain in his legs, even though he couldn't feel or move them any more. This is known as phantom pain.

'So these are the patients who are coming to me and I'm looking into their brains,' says Gustin. 'What is the reason why they can't get rid of their pain, or why their pain developed?'

To answer those questions, Gustin is studying two areas of the brain that appear to be quite significant: the **thalamus** and the **prefrontal cortex.*** The thalamus acts like a boom gate between the spinal cord and higher brain centres. When we have an acute injury, this border opens, allowing information to get through to the higher brain centres to tell us that we need to look after ourselves and heal. But after an acute injury has healed, this boom gate should close.

What Gustin and her team have seen in people with chronic pain is that this boom gate has *not* closed. 'So we found a decrease in volume in the thalamus, and this results in a decrease in a neurotransmitter called GABA, which usually dampens down signals. So in people with ongoing pain, the boom gate is always open and the signals remain strong.

'And we have seen a very similar mechanism in the prefrontal cortex, resulting in a decrease in GABA. This means that every emotion and every cognition is amplified. So people with ongoing pain, they anticipate pain with a lot of fear and they worry a lot of the time, and they can't dampen down these feelings because the prefrontal cortex has lost its ability to dampen down these thoughts.

'We recently revealed that not only is GABA reduced, but so is glutamate, a neurotransmitter that has the opposite function, stimulating nerve cells. Low levels of glutamate are linked to increased feelings of fear, **anxiety*** and negative thinking. Low levels of these two neurotransmitters may explain why people with chronic pain often experience mental health problems such as anxiety, **depression*** and suicidal tendencies.

'So we can change the way the brain functions and close the boom gate, and we can do that with **neurofeedback.*** We can change the way the cells talk to each other and we can actually rewrite the painful memories.' (I described the neurofeedback process in detail in Chapter 19.)

Medication is often used for chronic pain, but there are no drugs that directly target the neurotransmitters in the prefrontal cortex without causing side-effects to the rest of the body. As an alternative to pharmaceutical drugs, Gustin and her team have recently developed a program called Internet-delivered Dialectical Behavioural Therapy Skills Training (iDBT-Pain). It involves talking to a psychologist over Zoom about how to regulate the strong emotions associated with chronic pain. The results so far are promising.

Pain and body perception

Many surprising factors influence pain, including the way a person perceives their own body. This is one area that **Dr Tasha Stanton** is investigating. She is Associate Professor in Clinical Pain Neuroscience at the University of South Australia.

In Stanton's research she's found that people who have osteoarthritis in their hand perceive that hand to be significantly smaller than people *without* osteoarthritis do. She and her team have also seen problems with perception of touch. For example, people with knee osteoarthritis are not very good at localising where they are being touched on their painful knee. 'And some of the tests that I've done suggest that changes to the way that the brain processes that location-specific information may be underlying some of these problems,' she told me.

She continued: 'There's one theory that changes in perception of body size or of the location of touch are driven by the presence of chronic pain, and possibly changes in behaviour that occur when you do experience pain. For example, if your knee is really sore you often don't move it as much and so there is less sensory information being sent to the brain about that body part. But it doesn't explain everything. There is a possibility that in some people, these changes may have existed before the painful condition occurred.'

Stanton is using this knowledge about body perception and pain to develop possible new treatments. In an experiment to explore altered perceptions in people with painful knee osteoarthritis, she and her colleagues use visual illusions. 'We have participants wear video goggles and we provide them with a live feed of their knee. They watch their knee and leg move in real time, so they know it is their own leg … but then we change it in front of their eyes.

'One of the more potent visual illusions that we use is called the stretch illusion. Participants are looking down at their knee and suddenly they see it start to elongate, as if the joint is stretching out and being tractioned. At the same time we give a slight pull on the calf muscle towards the foot. So their brain is getting information from both vision and touch saying, "Your knee is stretching out big and long." And some people experience pain relief with this type of illusion.'

Stanton told me there's some evidence to suggest that information from one sense, such as touch or vision, can modulate the information that is coming from the **nociceptors** in our nervous system, which provide our brain with information about danger in the tissues. It's possible that in her stretch illusion experiment, the danger message is being changed by that visual input of the elongated knee, and this alters the pain experienced.

There are other experimental studies that show that if you apply a painful stimulus to someone and you change the size of that body part using visual illusion, it changes the way the brain evaluates the message from the painful stimulus, resulting in less pain. Stanton adds that neuroscientists still don't fully understand this pain-relieving feature.

She and her colleagues have also done some interesting work on our perceptions of back stiffness. 'We've all probably woken up with a stiff back at one point or a stiff joint in the morning and thought, "What did I do? Why is my joint so stiff?" And I wondered – is it actually stiff? I tested this idea and what I discovered was that our feelings of back stiffness don't relate at all to the actual stiffness of the back. And what's more, and I think very interesting, is that feelings of stiffness can be altered without changing the actual stiffness of the back, merely by pairing a particular sound with pressure to the back.'

When patients had pressure applied to their back when a sound like a creaky door was playing, they perceived greater pressure and felt more protective of their backs. When a gentler sound was played, they perceived the pressure to be lighter and didn't feel the need to protect their backs as much. Interestingly, there were no differences between healthy control subjects and people with back pain.

Stanton says, 'That is actually really exciting information, because it means that in people with pain, their ability to integrate and combine information from different senses is still intact, and that opens up the possibility that we can use different senses in treatment to be able to help improve clinical outcomes in people with pain.'

Limitations and prejudices

The potential for harnessing the power of our minds to heal our bodies is certainly exciting, and has already been proven to work in many cases – but be warned: it's by no means a universal miracle cure.

Jo Marchant told me that at the beginning of her research, 'I started out sceptical but open-minded. What really surprised me was the extent to which our thoughts and beliefs can create these biological changes ...

'But what the mind *can't* do is replace something that the body is missing. So for somebody with diabetes who is missing insulin, for example, the mind is not going to suddenly magic up insulin that it was incapable of making, or in cystic fibrosis where patients are missing a certain protein in the lung, the mind is not going to fix that.

'Also if the body is just overwhelmed by serious injury or infection or cancer, for example, the mind is not going to be able to suddenly get you out of that; there is no evidence that mental state can shrink a tumour, for example.

'Otherwise, the mind is very good; it plays a very important role in symptoms that we experience – so, pain, nausea, fatigue, itching, wheezing, all of those subjective things that affect our quality of life. Beyond that the mind can influence things like the cardiovascular system, the immune system, the digestive system. The limitation is we can't just *will* these changes to occur, these things aren't under conscious control. But by understanding when and how the brain is controlling these things we can come up with clever ways to influence them through other means, if you like.'

So, what has Marchant learnt from her research?

'I think I have found it quite empowering in terms of just those symptoms like pain, fatigue, nausea. I'm less ruled by them now, less afraid of them, if you like,' she says. 'I think I've just realised the power of changing your attitude to a symptom that you are experiencing, and just realising the role of the brain. So, not immediately thinking, "Oh, something is terribly physically wrong", but realising, "Well, actually, maybe I'm just worried or anxious, and if I go out and do something else or distract myself or just feel more positive about it, that could in itself create these biological changes and I'll get the release of endorphins in the brain that's going to help take that pain away."'

But mind–body science has a long way to go. 'I would love to see more research in the area generally. There's so little funding at the moment. The vast majority of clinical trials are into conventional physical drugs and intervention …

'I think in science and medicine generally there is still this idea of the mind–body split, and that physical measurable things are more valid for scientific study, if you like, than subjective elements like thoughts and emotions and beliefs. And that's not how neuroscientists would see it, they would see the two as completely intertwined, but I think there's still this hangover, this idea that the two are separate.

'When there are claims of the mind having an effect on the body, often the alternative healers describe that as some kind of mysterious power, the magical healing powers of the mind, and that turns a lot of scientists off. But I really don't see it as mysterious or magical, it's just biology.'

22

MUSIC IN MIND

Conversations with:

Daniel Levitin, James McGill Professor Emeritus of Psychology,
Neuroscience and Music, McGill University, Montreal

Dr Jeanette Tamplin, Senior Lecturer in Music Therapy,
University of Melbourne

Mark, subarachnoid haemorrhage survivor and music
rehabilitation patient

Louisa, Mark's partner

Annie, partial paraplegia and brain injury survivor and music
rehabilitation patient

Tobias Picker, composer, artistic director and pianist with
Tourette's syndrome

Dr Amee Baird, clinical neuropsychologist and Director, Newcastle
Neuropsychology

Dan Cohen, Founder and CEO, Right to Music; founder,
Music & Memory

Andrew Schulman, medical musician

The music in your brain

I invite you to play some of your favourite music in the background
while you read this chapter. It may just help you tap into the effect
that music can have!

Music has always been a central part of my emotional life – as
it has for most of us, across all cultures. Though I've never learnt

to play a musical instrument, listening to and appreciating music has been a personal passion from an early age.

In childhood, music touched me through the children's songs and nursery rhymes affectionately sung to me by my parents and grandparents. One of my most poignant memories is of sitting on my father's knee in the lounge room with some of his favourite music emanating from the highest-quality stereo system he could afford (or rather, *not* afford!). He would tap his long, strong fingers on my thighs in time with the music – which was most often jazz, blues or rock. These finger-tapping rhythms seemed to work their way into my soul and enhance the affectionate bond between us. No words required.

As I reached adolescence, the music I was listening to helped me establish my sense of identity. Unlike the parents of my peers, my dad would be playing the popular music of the '60s and '70s: the very same music my brother and I were listening to. We were never told to turn the music down; he would encourage us to keep it loud and crystal-clear.

As a young teenager, I discovered, to my delight, the Canadian singer-songwriter Joni Mitchell, who has remained one of my all-time favourite musical artists. Her exquisite voice, her expertise as a guitarist, her experimental approach to musical styles and the mind-blowing poetry in her lyrics all captivated me. Her music evoked a vast range of emotions – the joys, delights, fears and anxieties that defined that exciting and vulnerable time of my life.

You can imagine, then, how excited I was to unexpectedly share this deep appreciation of Mitchell's music with neuroscientist, musician and author **Daniel Levitin**, with whom I discussed the topic of music and the mind.

Levitin played rock music from childhood, and in his early adult life worked as a sound engineer, recording popular acts such as Steely Dan, Stevie Wonder, Santana and The Grateful Dead. He became so fascinated by why music so powerfully beguiles human beings that he began researching music and the brain. He's now

James McGill Professor Emeritus of Psychology, Neuroscience and Music at McGill University in Montreal. For his book *This Is Your Brain on Music*, he researched why and how music moves us.

How music moves us

'There's a particular part of the brain I've been studying for 15 years that holds a key to at least part of the question. That part of the brain is in the **prefrontal cortex*** and it goes by the rather inelegant but technical name of **Brodmann area 47**. Brodmann area 47 is responsible for helping us to form expectations about what's going to happen in the world ... And when we listen to music, Brodmann area 47 is exquisitely active, trying to predict what's going to come next in the music.'

Musicians manipulate our emotions by varying the timing, the timbre and the loudness of notes. Unpredictability in music is what attracts us to some types of music and not others. The brain is a detector of changing patterns, and if a piece of music is played too robotically, we become bored and disinterested. But if the music is *too* unpredictable, we become disorientated. The job of the performer or composer is to reward our expectations to a certain extent, while at the same time trying to vary the music enough to keep it interesting. It's the Brodmann area 47 in the prefrontal cortex that keeps track of all that.

Another part of the brain which has an important role in our experience of music is the **cerebellum**, a bundle of fibres at the back of the skull where the neck attaches to the back of the head. It's responsible for motor control, and for decades it was thought that its sole function was to help maintain regular timing when we walk or run, so our gait remains steady.

As Levitin told me, 'It's only in the last 10 or 15 years that we've gained an appreciation for how the cerebellum is involved in human emotion, and its role in music is particularly interesting. It is the brain's central timekeeper, and while any deviations in our timing when we walk or run are predicted by the prefrontal cortex, it's the cerebellum that's keeping track of whether things are happening with metronomic robotic regularity or not. And it has all these connections to the **frontal cortex** and the **limbic system**, the seats of human emotion, meaning that the cerebellum is acting as kind of a gateway to emotion.'

It's been shown through brain-scanning studies that whether we're listening, playing or composing, the stimulus of music activates more areas of the brain than anything else we know about.

Music and human evolution

The jury is still out in academic circles on whether music is an essential part of human evolution. The renowned cognitive scientist Steven Pinker calls music 'auditory cheesecake', meaning that it didn't evolve for any biological function, but is just an evolutionary accident.

But Levitin believes that the evolution of the brain has made music possible. 'One piece of evidence is that we have particular neurochemical systems that respond to music. When we listen to music, **oxytocin** is released; this is a *bonding* hormone. When we listen to sad music, **prolactin** is released, a *soothing* hormone. Playing music and listening to music can cause the release of **dopamine**, a *reward* hormone. So for neurochemical systems to be involved suggests some kind of evolution.

'Another piece of evidence that I find compelling is that music and language occupy very different neural circuits: they are separable. We've seen many patients now who, as a result of a stroke or a tumour or some other brain damage, lose one system and not the other. That is, the system of language or the system of music. So if music is really built on top of language, you wouldn't expect them to be separable in these ways.'

What determines our musical taste?

Levitin explains that musical tastes begin to form in the womb. By 20 weeks, the developing baby has a fully functioning auditory system. As the baby is growing in the womb, it experiences all the music that the mother is hearing and tries to make sense of it.

Familiarity is key, so by a year of age, babies show a preference for the music they heard in the womb.

'All Western music is based on the same musical rules. So when we talk about heavy metal or classical or jazz or country, mainstream popular music, we are talking about music that adheres to the same 12 tones, the same scale, the same rules, as distinct from, say, Indian music or Aboriginal Australian music or Arabian music or Chinese music – those have their own scales and their own rules. If you grow up listening to any of them, you internalise those rules, just as you internalise the rules of language, and you at least can understand that music and form preferences within that system.

'Then a third part of it is that in our early teen years we tend to gravitate towards music that helps to brand us and our social identity. At that age – 11, 12, 13, 14 – we realise for the first time in our lives that we don't have to just like what our parents like, we can have our own tastes and our own opinions. And part of that social identity is dressing in a particular way, liking certain books, certain movies and certain music.'

At this point in the conversation I was reminded of the music of my *own* teen years – and this was when Levitin and I indulged in fan-club-type banter, sharing our mutual appreciation of Joni Mitchell. I was keen to hear about an interview Levitin did with Mitchell in 1997: I'd always dreamt of interviewing her! She talked to him about when she was making the 1971 album *Blue*. She was at her most vulnerable emotionally, and shared so much of herself on the album. She told Levitin a couple of young fans came up to her and said, 'Before [the antidepressant] Prozac, there was you.' The music had a real therapeutic value for them.

Levitin found that a powerful statement: 'That's one of the beautiful paradoxes of art, is that through the intimately personal, art can become universal.'

When we're sad, we often feel alone and misunderstood, and then an artist comes along who seems to have been through similar low periods, and they are talking about it in an articulate, sensitive way, and suddenly we feel less alone, and we feel understood. This is the powerful therapeutic effect of a sad song.

Music as therapy

Even if we don't create or play music ourselves, just listening to a particular song at a particular time can really shift our perspective and have quite a therapeutic effect on your mood. We know that songs and lullabies can have beautiful and soothing effects on babies and young children, and the music of Joni Mitchell has the same therapeutic effect on me.

Many of us self-medicate with music, choosing the type of music we need to hear at a particular time. If we're feeling a bit down, music has the capacity to lift our spirits and energise us both physically and mentally. Or we may want to wallow in that feeling for a while with melancholic music.

Levitin believes music has huge potential to contribute to our health and wellbeing in therapeutic settings. He and McGill University colleague Mona Lisa Chanda outlined some of their findings in their literature review 'The neurochemistry of music: Evidence for health outcomes'.

They point to a series of studies where patients in a hospital who were about to undergo an operation were either given a sedative, or they were given soothing music to listen to. It was found that more people felt calm as a result of the music than they did as a result of the sedative. The music was also cheaper and had fewer side-effects.

'Another finding that we've seen replicated many, many times is that music can increase your physiological pain threshold. That means you feel less pain when you are listening to music that you like.'

Music therapy for rehabilitation

Dr Jeanette Tamplin is a Senior Lecturer in Music Therapy at the University of Melbourne. She specialises in the rehabilitation of people with neurological conditions such as traumatic brain injury, stroke, Parkinson's disease and **dementia**.*

She told me: 'Some of the particularly standout moments for me as a music therapist have been working with people who might be coming out of a coma and not verbal at all, just starting to come awake again, and the team are really trying to work on getting them to respond to communication over a period of weeks, and then having that moment in therapy when you are singing and they'll suddenly come out with a word, or you'll see them come alive, their eyes will open up. And that power of music to tap into some area of the brain that has connections with memory, has connections with emotion, has so many connections … Yes, it's really amazing to see those kinds of moments happen in a therapy session.'

Tamplin is a talented singer and musician and has also trained in the United States in the specialised field of **neurologic music therapy**. It focuses on the scientific application of music and how it affects our brains and physiology. It looks at the rehabilitation needs of the patient and applies a type of musical activity that will improve the patient's function, whether they're having physical or cognitive difficulties, or difficulties with language.

Tamplin says the largest body of research is in physical function or walking rehabilitation. Evidence shows that external rhythm has a strong influence on the way that we move. People who have had a stroke or a head injury that has affected one side of their body more than the other often have what's called a hemiplegic gait, where they walk unevenly. If they are played music with a strong rhythmic stimulus, Tamplin and her colleagues have often observed immediate improvement. This technique is called **rhythmic auditory stimulation**.

The Royal Talbot Rehabilitation Centre in Melbourne provides intensive rehabilitation to people with a wide range of disabilities, including acquired brain injury. They're using music therapy in some of their rehab programs.

When I spoke to Tamplin, they were doing two research projects. One of them looking at the effect of singing rhythm on people with speech and language disorders. Patients with aphasia, which is caused by damage to the brain, can no longer access language but they can still often use words through singing. Music therapists like Tamplin use a technique called **melodic intonation therapy**, which uses singing to take advantage of the undamaged right hemisphere of the brain by engaging areas that are capable of language. This helps people re-access speech.

Once they are articulating words in song, the next step is to gradually remove the melody, focus on the rhythm, and then return someone to normal speech.

Tamplin took me to meet **Mark** and his partner, **Louisa**, in the acquired brain injury ward. Mark was around 40 and had experienced a severe type of stroke called a subarachnoid haemorrhage. He was very lucky, because 80 per cent of people with his level of injury don't survive. He was in a wheelchair, and the damage caused problems with his language, cognition and memory. He had been at the Royal Talbot Rehabilitation Centre for around three months.

Tamplin explained that Mark had responded well to the music therapy she'd given him. In the first or second therapy session, she played the guitar and sang the song 'Knocking on Heaven's Door'. This was deliberately chosen, because it was a song that had been very familiar to him in his younger days. In early sessions he was able to come out with the word 'door', and over time he built on that and eventually could sing whole songs. His language outside of singing also improved. He remained confused, with a lot of ongoing cognitive and language issues, but he continually made progress.

I met Mark and Louisa, and listened as Mark sang 'Wish You Were Here', by Pink Floyd. 'I just love seeing him enjoy this,' Louisa told me. 'A lot of his work in terms of rehab is not fun, it's hard work, so I love seeing him enjoy himself.' She added, 'His vocabulary has really expanded, which is very special. Before we came here to the Talbot, Mark wasn't speaking at all.' His language improvement was made obvious by the cheeky and affectionate banter that went on between him and Louisa!

Tamplin then took me to visit another patient in the spinal injury unit, a girl called **Annie**. She'd had a road accident in which her car was completely written off, but she had no memory of it. Tamplin told me she had what's known as T12 incomplete paraplegia, a brain injury and some ongoing memory issues. She was participating in a study which was looking at the effects of songwriting on people who've have had a traumatic injury.

'Over six weeks they have twelve sessions and we write three songs over that time,' explained Tamplin. 'One is looking specifically at the past and their image of who they were as a person physically, socially, emotionally, as part of their family, academically, and then we write a song about that. And then we write a song about now – being in hospital, processing, adjusting to change, being a patient, having an injury. And then we write a final song looking towards the future and how they integrate their past with their present and then look towards the future. So it's really about an adjustment process using songwriting to help facilitate that.'

Tamplin told me Annie had written a song about her past and was halfway through a second one about the present. She described how together, she and Annie came up with the music to go with the lyrics.

'We kind of talked about what style of song and whether it was going to be happy or sad or mellow or upbeat, and kind of got a general idea, and then I just gave her some ideas, like, "What do you think about this?", and I'd pick something or I'd strum something and tried a few different chords. We'd have a line and then I'd sing it one way and sing it another way and say, "Do you like it if it goes up or if it goes down?"

'Annie is quite musical herself, she plays piano, so she has quite clear ideas about how she wants it to sound, so I just get her started, really, and then she goes, "No, I don't like that," or "Do it like that," or "Do it the way you did it the first time, as that was better," and that's kind of how it evolves, really.'

I asked Annie what the process had been like. 'Yes, it's interesting looking back at what I had,' Annie reflected. 'I don't really like what I was like before the accident, now that I'm looking back. I loved it at the time, but I think I'm a better person now … It's brought me closer to my family too.'

She also told me, 'I'm having a bit of trouble with my memory and concentration, and music therapy has definitely helped with both of those. Yes, so it has just kind of reminded me about things that I used to do … it was lovely, amazing.'

Remembering the past had given her hope that one day she would walk again. She was excited to show me that she could now flex her left leg and could feel it almost all the way down.

'Yes, so you've certainly got hope,' I said to her.

'Yes, *so* much hope.'

When I caught up with Annie recently, she said, 'It is interesting looking back at where my state of mind was all those years ago. I do think that music therapy was hugely important as part of my recovery.

'I have moved on past the idea of "walking again", which, while it can be a big part of early recovery, doesn't matter to me at all now. I am happy living independently in my own house and working at a community centre, which I love.'

Music, Tourette's syndrome and Oliver Sacks

Tobias Picker is an American composer, artistic director and pianist who found that he could use music to tame the symptoms of **Tourette's syndrome**, which causes uncontrollable physical and sometimes vocal twitching movements or tics.

'Music came very naturally to me,' Tobias told me. 'My earliest memories were improvising music at the piano. Finally my parents got me a piano when I was in third grade, so I guess I was eight.'

The teachers at school noticed a change in him after that. 'They seemed to think it made me happier, because I had seemed to them to be a rather unhappy child. They noticed that I couldn't sit still, but they didn't know at that time that I had Tourette's syndrome.' Nothing much was known about Tourette's in the early 1960s.

'I could vent my emotions through playing the piano, most especially my rage. I was very good at playing loud, fast, angry music … It was important to me that people knew, before I would allow anyone to come close to me, that I had this ability in which I could control something very powerful, a very big piano, the king of instruments, that I could wield my control over it. It was a compensation for my inability to control the tics, the twitching, as it was called. When I'd achieved a certain level of recognition and accomplishment as a composer, also after many, many years of therapy, a lot of that fell away and I became interested in meeting people just as myself.'

Then Tobias met renowned British neurologist Oliver Sacks. 'We met about 20 years ago through a mutual friend who had Parkinson's who introduced us. I gave a dinner party and invited them. I took a Valium before they arrived because I was nervous, and it had the effect of eliminating all of my tics, so Oliver's first impression of me was of somebody who, if anything, had a very negligible case of Tourette's syndrome. 'And it wasn't until a couple of years later, when I showed him a film that was taken of me at a rehearsal, that he saw my Tourette's in full bloom. He did see how my body was in constant motion, listening to a rehearsal of my second symphony. And he watched it over and over and asked me many questions about Tourette's and music.' Sacks was mesmerised by Tobias's story, and wrote about him in his 2007 book *Musicophilia: Tales of Music and the Brain.*

In 2010 Tobias completed his first ballet, inspired by Oliver Sacks's 1973 book *Awakenings.* The book is about victims of encephalitis lethargica, a condition that has some symptoms in common with Tourette's. 'I underlined all of the references to music that Oliver invented to describe their tics as motifs in conceiving the music,' says Tobias. 'It's astounding how much he relates the tics that these people are afflicted by in musical terms.'

Sacks was as admiring of the ballet as Tobias had been of the book. 'I played it for him before he saw it, and he was very moved by the music … He thought I'd written a masterpiece, he said. When he came to see the ballet I heard that he was speechless, but I think he was quite impressed to see this quite sophisticated choreography attempting to do something with his writings about the sufferers in *Awakenings*.'

Tobias has since written an operatic version of *Awakenings*, which premiered in June 2022.

Music and memory and aged care

Have you ever had that experience of hearing just a small snippet of music from your past and having a wealth of memories come flooding back?

'Many of us in the cognitive neuroscience field now believe that almost everything you've experienced gets encoded in memory,' says Daniel Levitin.

The difficulty is that all our memories compete when we try to access them. Music offers what's known as a retrieval cue, which helps us to distinguish particular memories because it often tends to be associated with a very specific time in our lives.

'And especially in the last 50 years, if you look at popular music, songs become hits for a summer and they are played constantly and then they disappear. That's exactly what you want in a retrieval cue, because it is uniquely associated with a time and place and with other memories of your life. So you hear that song that you haven't heard since you were 12 and instantly all of the other things that you were experiencing at the age of 12 come back in a flood. It wouldn't work with something like "God Save the Queen", or the national anthem or a song that you're hearing all the time; it works because it's so specific.'

He continues: 'Why would this be so? I think part of it is the nature of how music is represented in the brain. Music has all these different components: there's melody, there's rhythm, there's the lyrics, there's the tempo, the pitches, the harmony. And we now know that these are served by different neural networks, different regions of the brain. So there are these multiple reinforcing and redundant cues within memory. You may not be able to access all of them but if you can get just a few they trigger the others.

'I think the other thing is that in terms of lyrics, music is tightly structured, music has a certain tempo, and it has an accent structure … You add in the lyrics, which typically have to rhyme at the end of the line and the lyrics have to fit in with the melody. Even if you don't remember every single word, there aren't that many others that could fit. So your brain figures out what it must be and fills it in.'

Way back in Chapter 3 ('Memory'), I mentioned a concept called **prediction error,** which says the brain is constantly making predictions then correcting them whenever something different happens from what we expected. This predictive mechanism influences the way we hear and experience music. People tend to want a melody to return to the tonic, the first note of the melody's musical scale. Their brains predict that is what's going to happen. In many experiments where a person's brain activity is being measured and they are expecting the melody to go back to the tonic and it goes somewhere else, there's a big burst of activity in the brain. The repetition in music – of lyrics, of musical themes – also satisfies our brain's thirst for prediction.

Levitin suggests, too, that music is a better tool than language for arousing emotions: 'Language tends to be representational, it tends to be very concrete. Music is more metaphorical, and that ambiguity in it can allow it to mean different things to different people … Music that tends to be happy tends to have short notes and a lot of leaps in the melody and a fast tempo, and this is because it's modelling how we move when we are happy … When we are sad our body is a bit stooped, we move more slowly, we take shorter steps, and sad music mirrors all of this.'

Songs old and new

Levitin observes: 'We've now seen many, many patients in old age homes and nursing homes and hospitals, and they can forget what year it is, what city they're in, how old they are, their spouse's name, and yet they remember all these songs, typically from about the age of 10 to 14. They can sing along, they remember the melody and the words.'

Dr Amee Baird is a clinical neuropsychologist and studies the relationship between music and our memories. She told me that music stimulates wide regions of the brain that control emotion, memory and movement, and memory for familiar music activates the **temporal lobes** and **frontal lobes** of the brain. In conditions like **Alzheimer's disease,*** these parts of the brain seem to be relatively well preserved, and that's why familiar music has a powerful effect on people with that condition.

Baird was motivated to research this area because she read some inspiring case studies about musicians with Alzheimer's who, even in the late stages, could recall music by singing, playing their instrument or even composing.

'And I guess I could share some recent research that I've done with a 92-year-old lady who was not a musician but always loved music and always sang a lot, and her daughter contacted me and said, "She can sing along to new pop songs that come on the radio." And what was really exciting about this was that's a type of memory that is hard for people with dementia; to learn new things is something that's difficult. So I was really fascinated by this, and went to meet this lady in the nursing home and actually taught her a new song.

'Twenty-four hours later she could sing along to the new song that she'd heard that I'd taught her, and even one week later she could sing along with me to the new song that we'd taught her. And this was despite not knowing who I was. She had no idea who I was when I went to visit her a week later. She couldn't recall three words that I taught her after three minutes, but as soon as I started to sing the song, she could sing along with me.

'And I think this is really exciting, that there is this potential for music to be used as a memory aid, to maybe learn new information or be reminded of where you are or who your family members are.'

Playing music to people with **dementia*** can be hugely helpful for them. 'What it can do is it can transport these people quite quickly back to their personal memories. So it's a means of bringing them back to their personal past and a means of non-verbal communication.'

She adds: 'There have been studies of the positive effect of music on mood and in reducing **depression*** and **anxiety*** in people with dementia and other conditions.' Also, 'There have been studies of very strong effects on behaviour, in particular agitation in dementia, that music can reduce agitation. And also cognition, in that it can improve memory, as we've talked about, memory for the past in particular. But also in cognition it can improve people's language, expressive language. People start singing, or they produce more language in their responses after listening to personally preferred music.'

Dan Cohen is dedicated to promoting the use of favourite pieces of music by families and professionals caring for people with dementia and other cognitive difficulties. He is CEO and founder of the US organisation Right to Music, and I spoke to him about a 2014 documentary he featured in called *Alive Inside*. It followed several patients who have benefited from the Music & Memory program Cohen founded, which provides personal music playlists on an iPod to people in nursing homes with dementia.

One part of the doco went viral on social media and has become the most viewed video on dementia ever. It depicted a dementia patient called Henry being awakened from an inert, unresponsive state after being played his favourite Cab Calloway music on an iPod. When the music starts playing his face suddenly lights up. It's incredibly moving.

Cohen told me: 'I was in disbelief, I really was. I thought, "Wow, this is really cool." I looked around him and everybody was in disbelief – the staff, the fellow residents of the nursing home … So it's really something.'

Cohen told me about some other moving scenes he's witnessed. 'I'll often say in an aged care home, "Do you have someone that's just very disruptive or agitated?" "Yep, we have that person, with his hand he knocks away the food and drink from nurses, he is cursing at staff and it's very draining for everybody." "And so, well, what do we know about him?" "Well, not much." "What about his family?" "No, there's no family that we know of." But they did know he was a veteran, and so they made up a list of patriotic songs, and as soon as they put the headphones on him he snapped to attention and started humming to the music, and that was the end of his disruptive behaviours – bingo, problem solved. And so that was just a huge change.

'Another woman in Pennsylvania, she would always say the same four words, kind of repeating the same four words all the time. One day she saw a John Wayne movie and she'd mimic him saying, "Quick, get the gun!" But the problem was she'd say this any time of day or night, loud, so it bothered everybody. To her son, who was very attentive, we said, "Can we try the music with your mum?" "Oh, no, she never liked music, it won't work." But we prevailed and they tried it, and in two or three weeks she was singing the songs, she wasn't doing the iterative behaviours, wasn't disrupting everything, and everybody was thrilled.'

I was lucky enough to experience a trial of the Music & Memory program that was conducted in 2015 at an aged care facility in Sydney. I accompanied the staff while they sat down and chatted to one resident about her taste in music and what she used to listen to. The woman was quite open to several music suggestions and was happy for them to go away and compile a list for her.

On our way through the residence, we came across a man I'll call Tom, who was obviously enjoying his personal playlist. He was in his own world, swaying to the music and singing along at the top of his voice as Nat King Cole sang the classic song 'Daisy

Bell (Bicycle Built for Two)' through his earphones. Carers and other residents nearby joined in with him. It was such an uplifting moment to witness.

Then I was introduced to a man I'll call John, who was visiting his wife, whom I'll call Margaret, in the residence. She had late-stage dementia. John had informed the staff that Margaret was once a dancer and really liked music by The Carpenters. As John was sitting close to Margaret, tenderly stroking her hair, a staff member approached them with the iPod. She'd found out that one of Margaret's favourite songs was The Carpenters' 'Close to You'. John and Margaret were both given headphones so they could listen together.

When the music began playing, tears immediately welled up in John's eyes, while Margaret's response was subtler: her eyes opened wider, and she clasped her husband's hand. It was a beautiful and emotional moment. Everyone around them had tears rolling down their cheeks – including me. It was profound to witness how deeply that shared musical memory connected John and Margaret.

Let's hope that the Music & Memory approach can be used even more widely in aged care residences in the future.

The 'Medical Muso'

Andrew Schulman began playing the guitar as a child and has been a professional guitarist in New York for more than three decades. In 2009 his life changed forever, after he received a 99 per cent diagnosis of pancreatic cancer.

'And the miracle at first was that I got the 1 per cent. The mass was benign. It happens sometimes that way. But what followed that was that as I was being put onto the gurney to go from the operating room to the surgical intensive care unit, what we call the SICU, I went into anaphylactic shock. They'll never know exactly why, but it probably had to do with the blood transfusion.

'I arrived five minutes later after a mad dash with them rushing me on a gurney into the SICU and I was clinically dead, which means I was in cardiac arrest – no respiration, no blood pressure. But I was resuscitated.

'However, I was just massively ill from all kinds of complications. I was immediately put into a medically induced coma, and in the first three days of that coma, not a single doctor or nurse thought I would live. My wife didn't think I'd live either.

'It was in the middle of the third day when it was clear to everyone, especially to my wife, that I was probably within an hour or two of being gone permanently, that she had an epiphany.

'She was looking in her bag for her cell phone to call my mother and she saw my iPod, then she turned to the attending physician and said, "He loves music more than anything. Medicine is not working at this point, my voice isn't working to reach him, I think only music could reach him." And they agreed to try it.

'The earbuds went in, they didn't know what to play, they just hit the first track on the iPod. Luckily for me it was my ultimate favourite piece, the *St Matthew Passion* of Bach … it's very powerful music, very important music to me …

'And 30 minutes into the piece, all of a sudden I started stabilising, which I hadn't done in three days. I was terminally ill at that point, meaning I had terminal acidosis, lactic acid had built up in my tissues, and you don't survive that.

'But the acid started leaching out of my tissues. And by the way, it's an amazing story but it's backed up by the fact that it was in an ICU and doctors and nurses were there, and there's a chart for this showing all this stuff really happened.

'By that evening I was out of danger, and I never regressed, there were no more complications. They kept me in the coma for three more days … And for years in that ICU the nurses actually always talked about it as the St Matthew miracle.'

When Schulman emerged from the coma, his wife, Wendy, didn't tell him what had happened, because she thought it would be too much for him. A few days later, when Wendy wasn't there, he reached for his iPod to listen to some music. His favourite *St Matthew Passion* began playing again, 'and within seconds I started not just crying but weeping, weeping uncontrollably, and I had no idea why all of a sudden I was weeping like that. And it lasted for about 30 minutes, and it wouldn't stop, and I just was ... I didn't turn off the music, but I was just totally baffled.

'And Wendy comes back and she sees me crying like that, and she rushes over, she doesn't know what's going on, and I take out the earbuds and I say, "I don't know what's going on, I started playing the *St Matthew Passion* and I just started this weeping and grieving."

'And she nodded her head very quietly, and that was the moment to tell me, and then she told me. And we really figured out a number of things about that ... And the first thing I said was, "The body remembers." And there is actually a term that's called **embodied cognition**, that the mind is not just in the brain. Memory and cognition can be all through the body. And what it was is that my being, my brain and my full complete body, understood where I didn't consciously understand that that music had saved my life ... That's how I understood the power of it, I just didn't know it consciously, and then Wendy explained it. It was a very amazing moment.

'I knew that I had to give back, and that the only thing of value I had to give then was music ... All I wanted to do was work in critical care and use my professional musician skills, so I came up with the title Medical Musician, and it is now a new specialty ... And it's a professional musician with care training who specialises in intensive care units as part of the medical team.'

Andrew spent seven years as Medical Musician in the intensive care unit of Mount Sinai Beth Israel Hospital, walking through the wards and playing the patients their favourite music to help them in their healing and recovery.

In 2018 he founded the Medical Musician Initiative to train professional musicians to work within ICU medical teams, boosting the healing process of critically ill patients by using music as medicine.

Whether you are proficient in singing, composing or playing a musical instrument, or you simply appreciate listening to music, you will know that it can have a profound effect on you. It taps into our memory, emotions, our innate sense of body rhythms, it can improve cognitive function and be a powerful therapeutic healing tool for both body and mind. So why not take your cue from Andrew Schulman, and self-medicate with your favourite piece of music!

23

THE POWER OF
HUMAN CONNECTION

Conversations with:

Dr Michelle Lim, Senior Lecturer in Clinical Psychology and Head of the Social Health and Wellbeing (SHAW) Laboratory, Swinburne University of Technology, Melbourne

Steve Cole, Professor of Medicine and Psychiatry and Biobehavioral Sciences, University of California, Los Angeles

David Eagleman, Adjunct Professor, Department of Psychiatry, Stanford University, California

Nick Duerden, freelance writer and journalist

Dr Fiona Kerr, founder and CEO, The NeuroTech Institute, and Adjunct Senior Fellow, University of Adelaide

Marta Zaraska, science journalist and writer

Ralph Kelly, Executive Director and CEO, Stay Kind

Matthieu Ricard, Buddhist monk, humanitarian, author, translator and photographer

OneWave: My son Sam's story

Having heard and investigated many amazing personal stories about the mind, the brain and human behaviour, I've come to the conclusion that what matters to our mental wellbeing, above everything else, is genuine human connection. So I'd like to start this chapter by sharing a personal story of my own, about a community initiative I've watched grow over the past decade – a small-scale example that demonstrates the value of human connection. It's a movement that was co-founded by my son Sam Schumacher and his mate Grant Trebilco.

Sam fell in love with surfing as a teenager, and it has remained one of his passions to this day. In his early 20s he became good friends with fellow surfer Grant, who was living with the mental health challenges of **bipolar disorder.***

Sam and Grant both love what immersing themselves in the ocean does for them, physically and mentally. They got talking about that – aware that many of their peers also struggle with challenges to their mental health, often alone and in silence – and decided to offer others support and share their passion. That was how they started their community movement OneWave, a name that refers both to surfing, and to the idea that one wave is all it takes to get attention, connection and support from someone else when you're in need.

The movement began at their local beach, Bondi, in 2013. On Friday mornings at sunrise, they started holding gatherings on the sand, wearing their brightest fluorescent clothing, which led to the gatherings being called Fluro Fridays. Bright colours were encouraged because mental health concerns are too often invisible in society. On Fluro Fridays, participants shared personal stories or concerns, practised yoga, and taught each other the joys of surfing and the ocean.

The first Fluro Friday event held outside Bondi was later in 2013, in Byron Bay. The concept of Fluro Friday spread via word of mouth and beautiful colourful photographs posted on social media, and within two years there were 50 groups holding weekly Fluro Friday events at their local beaches. For OneWave's fourth birthday, 650 people lined up hand in hand along the shore of Bondi Beach, celebrating connection and support for mental health.

In 2018, during their visit to Australia to attend the Invictus Games, Prince Harry and Meghan Markle, the Duke and

Duchess of Sussex, asked to attend a Fluro Friday event at Bondi and joined the OneWave community there. As Sam's mother I was invited to join a small group to speak informally with the royal couple about the importance of mental health support. Who would have thought! This Fluro Friday royal visit was televised live across Australia and news articles appeared in many different languages around the world.

By 2020, the OneWave movement was represented in 50 countries, and that year's OneWave birthday saw 200 beaches join together in support of each other and the wider mental health issues in their neighbourhood.

The COVID-19 lockdown periods in 2021 prevented these communities from sharing mental health stories at in-person meet-ups. But because the need for mental health support was greater than ever at this time, the community stayed very active through virtual Fluro Friday sessions, where for the first time people from many different OneWave chapters met on Zoom. These monthly online meetings, dubbed 'Free the Funk', allowed people to share stories and worries, or just listen to and connect with others.

OneWave now delivers Free the Funk programs in primary and secondary schools and universities to raise awareness of mental health issues and share tips and tools that young people can use throughout their lives. Their message is that early prevention, connection and conversations save lives.

One amazing feature about this community organisation, is that apart from some sponsorship for the Free the Funk schools programs, it is managed by participants and volunteers.

Loneliness: A social pain

Most of us feel a degree of loneliness at some time in our lives. It's distressing and surrounded by stigma, so we find it difficult to talk about. Yet as many as one in four Australians has reported being lonely – even before the onset of the COVID-19 pandemic.

Dr Michelle Lim is a Senior Lecturer in Clinical Psychology and the Head of the Social Health and Wellbeing (SHAW) Laboratory at Swinburne University of Technology, Melbourne. I spoke to her after she had released the findings of the 2018 *Australian Loneliness Report*, a collaboration between Swinburne University and the Australian Psychological Society. As Scientific Chair of Ending Loneliness Together, she led the development and launch of the 2020 *Ending Loneliness Together in Australia* White Paper, and more recently in 2022, the *Strengthening Social Connection to Accelerate Social Recovery* White Paper.

She defines loneliness as a subjective sense of isolation. You can be around people, but still feel that you don't have a meaningful connection with others. You may feel like you're speaking the same language as people you interact with daily, but not really understanding them or relating to them fully.

Loneliness happens to us all, and Lim suggests that it serves a purpose: it's a warning sign to encourage us to live our lives within a wider social network.

People often experience loneliness during social transitions or challenges. One of those is during adolescence: young people are starting at new schools, making new friendship groups and establishing their identity. Another group who are vulnerable to loneliness are those over the age of 65. In both these phases of life we can feel a real loss of control.

That said, Lim notes that when asked directly, people over the age of 65 are the most likely to say they hardly ever feel lonely. This is where society's perceptions of loneliness don't always fit with reality.

Lim and her colleagues have found that loneliness leads to a lower quality of life and health. Higher levels of loneliness are associated with nausea, headaches, stomach problems and an increased risk of **depression*** and **social anxiety**.

Lim says there is a lot of research looking at how loneliness shares the same neural pathways as physical pain. When we feel pain, we have higher levels of cortisol, the primary stress hormone, which then affects our health. It's almost as if loneliness can be thought of as a *social* pain.

The scientific field of **social genomics** also shows how loneliness and social isolation can influence our physical health, and even affect us at a genetic level. **Professor Steven Cole**, from the School of Medicine at the University of California, Los Angeles, explained to me that social genomics looks at how our **genome** – the complete set of genetic information each of us needs to develop and function – 'plays a role in structuring both our health and normal development and how inadvertently it contributes to the production of disease and particularly of chronic illnesses'.

He conducted research in collaboration with the late John Cacioppo, Professor of Psychology at the University of Chicago and an expert on human loneliness. They analysed the genes of two groups of people; one group were lonely and the other were socially well connected.

'We looked across all 20,000 genes in the human genome,' Professor Cole says, 'and simply asked: "Are any genes functioning differently in these lonely people?" And we were really surprised by what we found. We found a big cluster of genes that were all involved in a very similar process. We actually found three clusters that were working systematically differently in lonely people's blood cells.'

Importantly, there are two types of loneliness: one is transient and doesn't result in genetic change, and the other is chronic and deeply entrenched in a person's life, potentially leading to changes in their genetic makeup. Cole explains: 'Over the last 10 years or so we've developed quite a lot of insight into how adversity influences the human genome, and how bad stuff gets into your body and affects your health. But I think the big question is, is there anything on the good side that can help protect us against those bad influences?'

He continues: 'I remain really enthusiastic about this question of how can we live the best lives for ourselves, both in terms of our psychological wellbeing and in terms of our biological health. So I think there's quite a lot more to learn about what we can do to protect bodies against adverse life circumstances.'

What we can do about loneliness

We could never have guessed the impact that the COVID-19 pandemic would have on us. We've all been affected in different ways but the need to stay physically distant from one another has highlighted the importance of human connection, empathy and kindness, both for our mental and physical health.

Between March and June 2021, researchers from the Matilda Centre for Research in Mental Health and Substance Use at the University of Sydney surveyed more than 1000 Australian adults aged between 18 and 85 about their experience of mental health since the start of the pandemic.

The conclusion was that a strained mental health system, financial stress and social disconnection had contributed to an increase in mental ill-health. Many who participated in the survey felt a lack of social and community connection were key concerns, most likely due to COVID-19 lockdowns and border controls. Some mentioned a restriction in spaces available for socialising, and others were concerned about a broad cultural shift away from valuing community. Loneliness is a growing issue around the world that's having a profound impact on our health and wellbeing.

However, there are ways to nurture our relationships and social networks, in order to help ourselves and others feel more socially connected. Lim says the first thing to understand is that feelings of loneliness are normal, and they don't mean you are less of a person. Loneliness can be managed. You don't always have to quickly make more close friends, but if you establish more meaningful social interactions, you can build relationships over time. Or you can look within your current network, and enhance and deepen the relationships you already have.

She also points out that there are probably many people outside your network who would like to establish new connections, so she recommends taking up opportunities to connect with people you can share common interests with.

It's also important to look out for people who may be lonely. Lim recommends connecting and being present with them without being judgmental. Listening is sometimes all that's required to help people with their loneliness in a meaningful way.

Social neuroscience

Back in Chapter 4 I introduced neuroscientist **David Eagleman**, from Stanford University in California. When we met he told me about a new scientific field called **social neuroscience**.

I asked him what it was all about. 'I think the main issue is that there is no meaningful way to mark the end of you and the beginning of those around you, because your neurons and those of everyone on the planet interplay in a super-organism ... The fact that we are an exquisitely social species allows us to bond and form all the fruits of our civilisation, which can really be understood as the deeds of a single super-organism. I know we think of ourselves as being different and unique and so on ... But in fact the effect of culture and your neighbourhoods around you means that everybody is influencing each other in this massive way and you are part of something bigger than you.'

A key feature of what connects us socially is **empathy** (something that will come up again later in this chapter). 'Neuroscientists have studied empathy for the last 15 years or so, and what it turns out to be is when you see somebody else getting hurt, the same networks that are in your brain, that care about *you* getting hurt, light up ... You are literally feeling somebody else's pain ... And it turns out that the more you care about that person, the stronger that empathic response is, and you are running the simulation of what it would be like if that were you.' (**Mirror-touch synaesthesia**, which I looked at in Chapter 8, is a very extreme version of this empathy that most of us experience.)

Eagleman told me, 'One of the experiments we've been doing in my lab – we're just about to publish this now – is this issue of what happens with **in-groups** and **out-groups**?' He explains: 'We show six hands on a screen, and then one of those hands gets picked by the computer and you see the hand get stabbed by a syringe needle. It's really awful-looking, and you have an empathic response to that. The networks in your brain that care about you being in pain light up.

'*Now* what we do is we label those six hands with different religious labels, so Christian, Jewish, Muslim, Hindu, Scientologist or atheist. And now the computer picks a hand and stabs it, and we are measuring what happens in your brain, and it turns out that if you happen to belong to that in-group you really care about it, you have a stronger empathic response than if the hand is labelled as a member of your out-group, of one of your out-groups, in this case there are five of them. And so in that case you have a smaller empathic response, just based on a one-word label.'

This idea of in- and out-groups may explain to some extent how human beings can be so cruel to one another, as we've seen in so many events throughout history.

Eaglemann says: 'This is a really important sort of thing because my hope is that the next generation will come to recognise things like propaganda and what makes certain people in your out-group – because as soon as you're told by your government or your parents or whatever that someone is in your out-group, you just care about them less. And so the hope is that the next generation will come to recognise these patterns of dehumanisation – literally dehumanisation, because the networks in your brain that care about someone as another human get dialled down – and will become more immune to this.'

He continues: 'I think that having an understanding of what's going on in the brain and showing this, showing this directly to the next generation is going to help so that they come to recognise these patterns.' In a world that desperately needs connection, breaking down the prejudices of the past can only be a good thing for all of us.

Look up and connect

Ninety-six-year-old Bob Lowe lost his wife to **dementia*** after spending five years diligently nursing her. Another five years later, he had not got over it. Though he had children who came to see him every now and again, they had their own lives, and he often found himself sitting at home surrounded by photographs of his wife, whom he missed and grieved for every day. For a while, at least, he felt very sorry for himself.

At one point he hurt his leg, and the nurse who was looking after him realised how lonely he was. She commented that he had a lovely voice, and perhaps he could volunteer to read newspapers for the blind.

Bob did as the nurse suggested. It really helped him pass the time and he very much enjoyed it. Eventually he found himself gainfully employed as a studio manager. He also started using a community call service, where volunteers make regular phone calls to elderly people just to chat and share news.

The more Bob talked, the better he felt, and, as time went by, though he still missed his wife, he found that his loneliness was decreasing. Then he began to help others, particularly men in similar circumstances, to manage their own loneliness. Tributes were paid to this inspiring loneliness campaigner after he died in November 2021, weeks short of his hundredth birthday.

* * *

I heard Bob's story from UK freelance journalist and writer **Nick Duerden**. He was asked to write about solutions to loneliness and social isolation after the senseless murder of British MP Jo Cox in 2016. Before her death, Jo Cox had set up a Commission on Loneliness.

When he began researching the topic, Duerden was shocked at the extent of loneliness in our society, and how it can affect everyone irrespective of age, race or class. His findings eventually led to his book *A Life Less Lonely: What We Can All Do to Lead More Connected, Kinder Lives.*

Duerden spent many months speaking to all sorts of people struggling with loneliness. He got the sense from everybody he talked to that people really *do* want to connect with each other; it's inherent in us to be part of a community.

For people like Bob, community call services and other forms of technology can be an invaluable tool for connection. Many of us discovered this during the COVID-19 pandemic, and for those of us with disabilities and ill-health who are unable to leave the house, technology can be nothing less than a lifesaver. It can provide a connection with the world that would otherwise be next to impossible.

But it's not just those who can't leave the home. With more and more of us living alone, as a society we are increasingly turning to our phones and technology in place of human contact.

Duerden points out how unhealthy it is to live exclusively on our phones and rarely look up at the world around us.

Dr Fiona Kerr agrees. She's a cognitive neuroscientist, complex systems engineer, anthropologist and psychologist, and founder and CEO of The NeuroTech Institute, exploring how humans interact directly, and via technology. She's had a consuming interest in the power of human connectivity for more than 35 years.

Kerr asked me what is the first thing that I do when I queue up for my morning coffee: 'Do you look up and connect with someone, maybe have a conversation, do you daydream, or do you dive into your mobile phone world?' I have to admit I'm often guilty of the latter!

Kerr says the option you choose will make a big difference. The science of human interaction tells us that the way we connect with each other has an impact on our brain and our body. Face-to-face connection changes us and the society we live in.

Interacting with someone in the line, even if you don't know them, starts up a chemical cocktail, where various parts of the brain associated with social and emotional functions put out feel-good hormones such as oxytocin, dopamine and vasopressin. Kerr refers to spindle and mirror neurons, which deal with brain connection and recognising behaviour in others, and hormones that make you feel safe and content. You feel like you're connected to other human beings.

'If you want to daydream or just wait and stare out of the window,' she says, 'again you allow this beautiful abstraction to occur. So you let your brain go. And when we don't hamper our brain by either thinking of a task or by distracting it in other ways, then it starts to make connections, maybe about a problem you're thinking about or something creative, and when you're looking up and out, then there's a number of different parts of your brain that start to turn on all at once which don't normally turn on, so you can actually watch that daydreaming mode click in. And your brain is very, very active and it's really good for it. And again, you can get a number of nice chemicals from that.' (This tallies with what Professor Muireann Irish says about daydreaming in Chapter 6.)

'What happens if we pick our phone up is neither of those. We tend to turn off abstraction and we tend to give a signal that we are not going to communicate, we're busy. And so either we pick something up which is distracting, like a game, or we decide to do something useful like emails or think about a problem.

'And one of the interesting things about it is if you were to either interact or to daydream, your brain is probably much more efficient at dealing with the problem that you're trying to think about than if you try and do it quickly in an email or process the situation on a screen, because what that does is pull your brain to task immediately and it stops a lot of the cross-connecting which gives you that "A-ha!" moment that you get when you are actually looking up.'

How human connection changes our brains

Kerr's research is all about measuring the chemical and biological changes that happen when we look up and connect with others: 'We all know that if there's an issue or if there's an emotional situation, positive or negative, we want to look directly at someone else. So it's why we still do face-to-face meetings … And we know from Finnish research that if you look at someone's face over a screen, there are certain parts of our brain that never turn on, whereas if you look at them face to face those bits *do* turn on.'

Kerr says when you meet and interact with someone, especially if you are likely to see them again, you go away with what she calls a 'neural calling card' of that person. During the interaction, neurons in the brain create a small new physical network for the person you've just met – so that when you talk to them again, your brain physically switches on that network and creates more electrochemical activity when connecting with that person.

Kerr also believes eye contact is one of the really powerful ways in which humans connect with each other: 'If you are in an emotional state, what happens with humans is we tend to look at the eyes of someone we trust, and especially if it's someone that we have a warm relationship with, but you'll actually look at anybody rather than no one, because we have a huge number of oxytocin receptors in and around our eyes that are rewarded by a direct connecting gaze. It increases electrochemical activity in our brains and our bodies, and we start to change our biochemistry. One of the outcomes is to alter the way we, or others, then deal with a stressful situation, both emotionally and physically. Oxytocin assists the immune system, and a direct, warm gaze can not only calm an anxious patient but positively impact healing. Add a warm connecting voice and a touch on the hand or arm, and the neurochemical impact is profound, and remarkably efficient.'

How technology is changing us

Human beings have always related to each other face to face and connected through eye contact, speech and physical touch, but there is now concern that new technologies like mobile phones are significantly changing our habits of interaction.

'We actually know a lot about the fact that we get addicted, we get dopamine spikes,' Kerr says. 'That's the sort of thing that draws us back to the phone all the time. It's very insistent, so it's quite different from any other technological advance that we've had before.'

What the technology does, she says, 'is make us feel conflicted, so it's very intimate, yet at the same time we don't have intimacy if we are just over a screen. How do you connect physically? You connect through neural synchronisation … We connect through voice, which is resonance and we actually pick it up. We connect through physical space, so we resonate and we give off a number of chemicals between our bodies.'

She observes: 'One of the things that I worry about sometimes is watching a young parent feeding a baby and texting at the same time – now they've got a title for it, "brexting", which is breastfeeding and texting at the same time.

'When a baby looks at you, their focal point is perfect for you staring into their eyes while they are feeding. And there's a huge number of things going on in their brain: their pleasure centres are going off, they're getting glucose so they're actually building in their brain, a huge amount, the connection with you, the resonance of your voice. They are building lots and lots of connections in there. And when you look away, a number of these activities stop.'

The problems continue as children get older: 'If you have the eight-year-old girl who is trying to talk to you and you are on your phone going, "Just a minute, hang on", what you're actually saying is "You're not as important; that can wait." And they've found that things like the feeling of self-respect drops, especially in girls. So those sorts of things really matter when the child is growing. And validation for our worth and existence is just as important for those who are at the other end of their lives, in maintaining wellbeing and even slowing various negative neurobiological impacts of ageing.'

it make a difference to the people they were trying to help, but it helped *them* as well. People are inherently good, Duerden thinks, and perhaps we need to be reminded from time to time.

By slowly reaching out, we'll start to build connections, and crucially, we'll learn that we are not alone.

The secret to a long life

Science journalist **Marta Zaraska** has long had an interest in the factors that help us live to a ripe old age. While researching a book on the topic, she'd been working on the traditional science of longevity, which focuses on diet and exercise – but she stumbled upon an increasing number of studies that were pointing in a *very different* direction.

These studies showed that the so-called 'soft drivers' of longevity, such as empathy, kindness and relationships, may matter even more to our health than nutrition and exercise. After reading hundreds of research papers and talking to dozens of scientists, Zaraska concluded that the things that matter most are how we live our lives emotionally and socially, the connection between the mind and the body, and having a sense of purpose in our lives.

'So definitely the more super-centenarians you meet, the more secrets they will tell you,' she says. 'But I did meet quite a few centenarians in Japan when I was travelling there for research, and one of the very common themes there, one that actually scientists agree on as well, is that a very important thing is to have a purpose in life, or what the Japanese call *ikigai* or a reason for living, and they consider it so important to health that even the Health Ministry of Japan recognises *ikigai* as one of the most important health drivers. Just like they tell people not to smoke, they tell people to exercise, they also tell people to find their *ikigai* … And this is the thing that really shows up in Japan a lot, and there is research confirming that indeed having purpose in life is a very important thing for your health and your longevity.'

Just as important is social connection. 'So one very interesting adventure, you could say, I had when researching this book was when I travelled to Oxford and I met scientists at the University of Oxford working there on the connection between our gut microbes and our social lives and our health and how all this influences each other all in one big circle. So I actually found myself in the forests surrounding Oxford, catching mice with other scientists to check their gut microbes and their social connections ...

'And the interesting outcome of all this research was that the more social the mice were and the more diverse relationships they had, the healthier their microbiomes – the healthy gut microbes, which also influence their general health and behaviour. And we know that the same thing is actually happening with humans as well; we know that people exchange microbes, for example, when we are doing contact sports – you will be exchanging your gut microbes through skin contact with other people. The same with your family and friends, when you hug them you exchange gut microbes.

'And the mice study tells us that most likely when we have diverse relationships – so with very different types of people – our gut microbiome *also* becomes more diverse, which is generally a good thing for our health.'

Zaraska adds: 'Volunteering is definitely one of the best things you can do for your health. It's actually so good for you, in fact, that volunteers spend significantly less time in hospitals, for example. And you can even see the effects on volunteering directly also on the inflammatory markers ... That people who volunteer have significantly lower levels of one inflammatory marker called interleukin-6, and this is a very good thing for you because it lowers your risk of premature death.'

So, how do we cultivate that sort of kindness towards others? Zaraska has discovered that 'kindness is definitely connected to the level of your empathy, and ... the truth is that empathy can be really trained. So there are organisations that I witnessed at work, for example one called Roots of Empathy, which is a Canadian charity that goes to schools across the planet and teaches children empathy. They have a very interesting process where they bring a newborn baby on a regular basis to the classroom and children are interacting with the baby, trying to understand its emotions and see it grow. And there is research confirming that it really helps boost the children's empathy.

'And we know from other research as well, for example on medical students, that you can really train your empathy, generally by trying to look at the world from other people's perspectives; even reading books or watching movies that are emotional really helps us improve our empathy as well. So this is also like a muscle that you can train. It's not that if you were born not particularly empathetic then you are doomed to stay this way; it's just not the way it is.'

So many of us have our health priorities all wrong. 'We spend so much time reading about diets and looking for new exercise regimes and uploading new apps to keep our muscles the strongest possible and looking slim and so on and so on, so much time and effort goes into it, and how much effort do you put into your social relationships, into the quality of your friendships, into your empathy, into your marriage, you know, as all these things are possibly even more important to your health.'

Simple things can make *such* an impact: 'Just looking at things differently, so realising exactly that my connections to my community, the volunteering, these are all health behaviours, not something apart from my diet or exercise. Even things like really small acts of kindness matter. So if you let somebody pass in traffic ahead of you, this is health behaviour, just like eating an apple.

'So if you think about this in these terms it really changes the perspective.'

Stay Kind

Ralph Kelly is director and CEO of Stay Kind, formerly the Thomas Kelly Youth Foundation. It's a movement that promotes

and encourages more kindness in everyday life. Kelly was inspired
to set up this foundation in 2012 after he and his family endured
a terrible tragedy.

'We lost our eldest son Thomas, who was 18 at the time, and
going out for his first night in Sydney. He had met with some
friends in the city and they had dinner together … and shared
taxis from the city up to Kings Cross. So Thomas's taxi dropped
him in Victoria Street. He was literally … he had taken his first
steps on the pavement when his attacker stepped off the wall and
punched him. At 10.07, which was literally two minutes after he
got out of the taxi, he was brain-dead. So the impact of the force,
which he didn't see coming, and then the impact of the pavement
was enough to kill him immediately, although he was on life
support for two days at St Vincent's Hospital.'

Kelly described his reaction to me. 'I kind of went into shock,
to be honest with you. I couldn't work, I couldn't function, I
emotionally shut down on every single level and spent two
months just sitting in a room wondering what happened. All my
friends kept on calling me and sending me emails, but I'd never
answer them. And then I started to feel terrible.

'So this friend of mine who really badgered me every day, I
answered his call and he said, "Ralph, we need to do something."
And I said, "Well, I don't know what to do." He goes, "Let's start
having meetings with people who understand what's happening
in the community at Kings Cross." So we started to meet with up
to five people every day. It kind of took over our lives.'

After months of intensive work with community leaders, Kelly
started the Thomas Kelly Youth Foundation to provide help and
support for youth across the community.

Then, just four years after Thomas died, further heartbreak
struck the Kelly family. Their youngest son, Stuart, took his own
life at the age of 18.

Kelly told me: 'Losing Stuart was just awful. I think if you
lose a child to homicide and then to suicide, you've kind of won
the double jackpot, the worst jackpot you can ever win. It was

a time where I started to reflect on do we just shut down the foundation, because we have a daughter, Madeleine, and just stop? And then we had so many people relying on services that we were providing in the city and our volunteers had become like family, these are young people and people in their 70s who come out on the streets with us all night from 10pm until 4am, and I kind of thought about it, and I thought, "We can't let all these people down who haven't let *us* down."

'But we wanted to change the name of the foundation, because it was in Thomas's name, and in reflection it was affecting Madeleine … we called it Stay Kind, because if you get up every day and you know that you're going to stay kind and stay in the moment, you're going to get out of bed, you know that you're going to help other people …

'And so when we looked at some global research we saw that … we should try to teach people, children and adults, respect and empathy. But if you don't have kindness as the underlying value, you can't even teach respect and empathy, because if people don't have kindness they will never have the other two values. So really kindness is the underlying value in every single part of our life.'

Kelly says to people if we all did one kind act every day, in one month alone there would be 770 million acts of kindness in Australia.

'And it's good for ourselves when we do acts of kindness because it makes us feel better, but it will also make the other person feel better and the receiver will then pay that forward at least three to five times. So it's not about our foundation, it's not about our families, it's not about Tom or Stu, it's about trying to make Australia a fantastic country again.'

I wondered how Ralph and Kathy Kelly were able to carry on their work and 'stay kind' after such intense personal sadness.

'If you had kindness,' Kelly replied, 'then both Thomas and Stuart would still be alive today. I've got very good friends whose children have been horribly bullied as they go through school, and I think we all recognise that to some degree, but all of us choose to continue on with our lives and not do anything about it. So it takes courage to be kind. And what we are saying is that kindness is simple, this is not difficult to do, it's very straightforward, it's very simple. It can be saying hello, opening a door, giving up a seat on the bus – really everything in your life, you can change everything, you can change other people's days, you can actually save people's lives.'

Cultivating altruism

As a young French biochemist, **Matthieu Ricard** rejected his promising academic career and headed to the Himalayas to become a Buddhist monk. 'I ... had to do a post-doc in America, as my boss wanted me to. I told him I wanted to do a post-doc in the Himalayas, and I'm still doing it 50 years later.'

Ricard has been practising and writing about altruism for many years, and teaches meditative techniques that nurture compassion. He and his friends began a humanitarian organisation called Karuna-Shechen, which uses compassion in action to support the underprivileged populations of India, Nepal and Tibet. I spoke to him when he was in Australia for the 2017 Happiness & Its Causes conference. My conversation with this relaxed, warm and sometimes quite funny character was a delight.

Ricard believes that each of us can learn to be more altruistic: 'The individual has to desire changing and spend the time to change. So that's exactly what mind training is about and with brain plasticity, as well as contemplative science, you can change if you train. You can learn to play badminton, you can learn to be more compassionate and altruistic; it's the same process, it's training, and your brain will change, you will change, it will change functionally, structurally, you will be a different person.'

Ricard has been the French interpreter for the Dalai Lama since 1989. In 2000, the Dalai Lama, who is very interested in science, invited him to participate in research on the effect of meditation and mind-training on the brain.

'One of the first things they found was especially when we engaged in compassion meditation there was a level of activation of certain areas of the brain connected with positive affect, also with empathy and with prosocial behaviour, with parental care, for instance ... Especially also with EEG [electroencephalogram], a level of activation of gamma frequencies which are related to coherence in the brain that were much, much higher than anything that was recorded previously in neuroscience. Then with the brain imaging it was found that meditators activate ... areas of the brain at will that are connected either to attention, to empathy, to compassion and to what we call open presence, which is a very vast open state where you notice everything.'

But Ricard feels that training yourself to practise altruism does more than change your brain. It's also the key to a more positive and sustainable future for our entire planet.

He explains, 'If you see the bigger picture, selfishness will not do the job ... We all want to build a better world, but we have to sit at the same table and have a unifying concept ... The American comic Groucho Marx said, "Why should I care for future generations, what did they do for me?" The problem is some people, you know what I mean, are saying the same thing seriously, which is a drama.

'So therefore I think [that with] more consideration for others, which is the definition of altruism also, you get a positive caring economics where you really care for poverty in the midst of plenty, and common good. Then you get more social justice, you can address inequalities. And then you have consideration for future generations that will be there, that will say you knew and yet you did nothing. So altruism becomes not just like a lofty, utopian, nice idea that [if] you have the luxury, [if] you can afford [to], if everything goes well, then you can be altruistic. It is the most pragmatic answer ...'

It can be difficult to believe in altruism, though, when we're bombarded through the media by stories of hateful, violent and selfish acts.

I put this to Ricard. He responded, 'One reassuring thing is what I try to call the banality of goodness. We are flooded by *bad* news, but there's plenty of *good* news. People under the poverty line went from 1.5 billion to 700 million. Mortality of children has halved over 20 years. Violence has diminished over the last centuries. You know, we went from 100 homicides per 100,000 in Europe to one. One hundred times less. All these things happening, a better world, yes, we have a better world.

'And then we can assume that most of the time most of the 8 billion human beings behave decently to each other, and that's normal, that's why we are so shocked when someone has committed barbarian acts and everyone immediately pays attention to that, and you find that on all the TVs of the world because it is aberrant, because it is deviant, because it's barbarian. We are equipped by evolution to pay attention to a potential deviant that is dangerous. We don't pay the same attention to everyone behaving nicely.

'So that distorts our perception of the world. The wicked world syndrome ... so we can say, "OK, fine, look at the world, it is improving much more than what we think, but it could still be better ... So we need to be more altruistic to consider future generations and change our behaviour. How to do that? There's two ways. You cannot just change institutions because then it becomes a totalitarian system: "We know what you should do and think and behave and then if you don't like it we will have to eliminate you." That doesn't work, we went through that. So you have to begin at the level of the individual ...

'Now, when a critical mass of individuals change their behaviour, their goal, their intention, their outlook, their way of being, the way they see each other, not as everyone for themselves but we are together in the same boat, a critical mass, then it tips over a change of cultures. And evolution of culture is a Darwinian process, but it's much faster than genes, as some researchers have shown. So instead of having to wait for 100,000 years for having a more altruistic gene, you can have a more altruistic culture selected by evolution.

'Why? It's quite simple. If you have a group of altruistic people, they usually like to work together by nature, they are altruistic, they are cooperative, they love working with each other … So as a community, which is not a real community, it's a bunch of selfish individuals, if the altruistic get their act together they have a strong advantage. So in fact at our time, at this juncture the trait that is best for our survival is cooperation, it's more altruism; it is not reckless competition, that's the quickest way to eliminate ourselves out of the planet.

He continues: 'Now, if culture changed, the next generation will be raised in that culture, you should respect others, you should not be racist, you should not be sexist, no gender discrimination, you should not be homophobic … So if you are raised with that, your genes don't change but your brain will be definitely configured in a different way.

'Individuals and groups and culture shape each other like two blades of a knife that are sharpening each other. Individuals make culture change, the next generation changes even further, and then they change the institution, then the whole thing takes on a different picture.'

* * *

Matthieu Ricard's observations instilled in me a sense of great calm, as well as optimism for humankind. Over many years I've had the privilege of hearing from and spending time with a wide range of people willing to talk about their mental health experiences, as well as psychiatrists, carers, and pioneering scientists at the forefront of brain research. All I've heard, seen and read convinces me that Ricard is right and that our growing understanding of the brain and the myriad ways it impacts on our behaviour and experience is not only allowing us to cure illnesses and improve quality of life for millions of people, but also giving rise to greater empathy, tolerance and kindness – which in turn points towards a brighter future for us all.

It has been a pleasure to share what I have learnt with you.

GLOSSARY

Alters

Two or more distinct personalities experienced by a person with **dissociative identity disorder (DID).*** Sometimes referred to as 'parts'.

Alzheimer's disease

A disease that affects 70 per cent of **dementia*** patients, marked by a buildup of amyloid proteins and malfunctioning tau proteins.

Amygdala

A part of the brain's limbic (feeling) system that specialises in handling intense emotions such as anxiety, anger and fear. It plays a key role in our reaction to pain and helps trigger the fight or flight response – often inappropriately, resulting in chronic **anxiety*** and stress. It is also one of the places where our long-term, explicit (conscious) memories are stored.

Anxiety

A natural response to physical threat in our environment – part of the fight or flight reaction triggered by our sympathetic nervous system. This reaction can be inappropriately activated by *perceived* threats – fear of social situations, performance, health issues – and when the anxiety is serious and prolonged, it can become a mental illness. It's the most prevalent mental health problem in Australia.

Asperger's syndrome
A condition involving difficulties with social interaction and repetitive patterns of behaviour. It's now classified by **DSM-5*** as being at the milder end of the **autism*** spectrum, though many people diagnosed before DSM-5 still see themselves as having Asperger's.

Auditory hallucinations
See **Voice-hearing**

Autism spectrum disorder (ASD)
A condition that is seen as a spectrum because of the huge variety in symptoms and severity, both between individuals and over each individual's lifetime. It's broadly characterised by social and communication difficulties, repetitive behaviours, limited interests and sensory sensitivity. The spectrum encompasses children who are unable to speak and require constant support, all the way up to high-functioning, gifted adults with minor social awkwardness.

Bipolar disorder
A disorder involving extreme mood swings, from depressive lows to manic highs.

Borderline personality disorder (BPD)
A serious disorder frequently caused by experiences of trauma and characterised by extreme and unstable emotions. BPD sufferers often struggle with their identity, exhibit impulsive behaviours and have difficulty maintaining stable relationships. Drug and alcohol abuse and self-harm are common 'self-medication' strategies, and around 10 per cent of BPD sufferers take their own lives.

Cerebral cortex
A deeply folded and wrinkled layer of grey matter on the outside of the brain, where huge quantities of information are processed.

Charles Bonnet syndrome (CBS)
A condition experienced by those with vision impairment from a lesion or abnormality anywhere along the visual pathway, involving

hallucinations that are entirely visual. It's estimated that 25 per cent of people with this kind of visual impairment have had at least one CBS episode. The precise cause is unknown.

Corpus callosum
A thick bundle of nerve fibres connecting the two hemispheres of the brain, allowing information to be transferred from one side to the other. This is important for any brain functions carried out by both sides of the body, and for other functions where one side of the brain needs to dominate.

Deep brain stimulation
A procedure that involves drilling holes into the skull and inserting tiny electrodes into the deepest parts of the brain to stimulate neurons and their networks. The electricity can be turned up and down to modulate how different parts of the brain operate. It is currently used to relieve the symptoms of Parkinson's disease, epilepsy and Tourette's syndrome, but there is also potential for it to be used on patients with mental health disorders such as **depression,*** addiction, obsessive-compulsive disorder and anorexia nervosa.

Default mode network
One of three separate brain networks, consisting of multiple brain areas working together to carry out complex functions. The default mode network is activated when we're relaxed and our mind is wandering. This network is responsible for remembering the past, imagining the future, self-reflection, creative thinking and daydreaming. The other two networks are the **executive control network,** for focusing and managing multiple thoughts, and the **salience network,** for filtering salient information.

Dementia
A collective term for 100 to 200 different conditions, usually indicated by a decline in two or more intellectual abilities such as memory, thinking, judgment, language, behaviour and life skills. It's experienced by around 10 per cent of people over 65, though early-onset dementia can occur in people as young as their 30s.

Depression

A term for many different illnesses that occur when our brain circuits have trouble communicating with each other. Depression is more than just being sad, and each person experiences it differently, and it has a cumulative effect. It's caused by a combination of genetic and environmental factors (including stressful events in our lives).

Dissociative identity disorder (DID)

A rare psychological condition previously known as **multiple personality disorder**, involving a dissociation or involuntary escape from reality, caused by extreme or sustained trauma. The sufferer's identity becomes fragmented into two or more separate personalities, known as **alters.*** Some people refer to them as 'parts.'

DSM-5

The fifth edition of the *Diagnostic and Statistical Manual of Mental Disorders*, produced by the American Psychiatric Association (ASA) in 2013. The DSM is often called 'the psychiatrists' bible'. A text revision, known as DSM-5-TR, was published in March 2022.

Ecstasy
See **MDMA**

Epigenetics

The study of how environmental and behavioural influences can change the way our genes are expressed.

Face blindness
See **Prosopagnosia**

Functional plasticity

The transferral of functions from a damaged brain area to other, healthy areas.

Highly superior autobiographical memory (HSAM)

Also known as **hyperthymesia**. A rare and selective type of memory ability that allows someone to recall an exceptionally high number of autobiographical and episodic (day-to-day) memories, and frequently the dates when they occurred.

Hippocampus

Part of the **limbic system,** the feeling part of the brain, and one of the places where long-term, explicit (conscious) memories are stored, particularly explicit memories that happened at a particular time and place.

Intergenerational trauma

Trauma that is passed down through generations, commonly experienced by Australian Aboriginal and Torres Strait Islander people, particularly the descendants of the Stolen Generations and those affected by the historical trauma of colonisation.

Locked-in syndrome

A rare condition in which the sufferer appears to be in a vegetative state, but is in fact fully conscious, often only able to move their eyes. It usually occurs after damage to the ventral pons, part of the brainstem, and can also result from tumours, traumatic brain injury or stroke. Sometimes it has no known cause.

LSD (lysergic acid diethylamide)

A synthetic psychedelic drug used in the treatment of **depression.*** Psychedelic drugs switch off the **default mode network,*** which controls brain function, and this allows the brain to behave in new ways.

Lucid dreaming

A dream in which you're aware that you're dreaming, like a dream within a dream. It happens during REM sleep and commonly lasts for two or three minutes, but sometimes for up to an hour.

MDMA (3,4-methylenedioxymethamphetamine)

Also known as ecstasy, this drug used in the treatment of **post-traumatic stress disorder.*** It reduces activity in the **amygdala,*** where we process fear, and connects the amygdala with the

hippocampus,* where long-term memories are stored. Memory for fine details is enhanced, while previous fearful emotions are reduced, helping patients face their trauma and accept it as something that happened to them in the past.

Mind–body connection
The idea that our minds can play a part in healing our bodies, once treated with suspicion by scientists, but now increasingly backed up by research.

Mindfulness
A meditation technique that involves paying non-judgmental attention to the present moment, increasingly used for managing mental health.

Multiple personality disorder
See **Dissociative identity disorder**

Neuroaesthetics
A new field of research seeking to discover what is going on in our brains when we admire a piece of art. It uses **neurofeedback*** to study the effects of visual art on our brain circuits.

Neurodiversity
A growing movement based on the idea that we are all born neurologically different, and those differences are something to celebrate. Neurodiversity grew out of the autism community but has now expanded to embrace a whole range of conditions.

Neurofeedback
A form of **biofeedback**, which uses computer software and sensors connected to the body to measure blood pressure, brainwaves and body temperature. This information is immediately fed back to the client so they can use it to change the way they are thinking and feeling and become more resilient. Neurofeedback is used to help trauma patients avoid states of dissociation and agitation, so they can function normally in their daily lives.

Neuroplasticity

The brain's ability to change itself by strengthening some neural pathways and pruning away others (called **synaptic pruning**). *See also* **Functional plasticity** and **Structural plasticity**.

Obstructive sleep apnoea (OSA)

A condition in which the throat narrows, or closes over completely, up to 100 times per hour during sleep, which makes breathing become very difficult. It affects over 1 billion adults worldwide.

Parasomnia

An unusual behaviour experienced during REM sleep, non-REM sleep, or the period between sleep and wakefulness. These behaviours include hallucinations, sleep paralysis, narcolepsy, cataplexy and acting out our dreams.

Placebo effect

A measurable improvement in a person's physical or mental health after taking a treatment they believe will be effective, even though the treatment is actually a fake. It is unethical for doctors to tell patients they are getting effective drugs when they are not, but there is research showing that 'honest' placebos – where the person is told they are getting a placebo – are still effective. Placebos can trigger measurable biological changes in the brain and the body that are similar to those caused by drugs, providing powerful evidence of the **mind–body connection.***

Post-traumatic stress disorder (PTSD)

A stress-related disorder resulting from exposure to an intense single experience of trauma, or repeated traumatic experiences over a long period of time. These can include child neglect, abuse and sexual assault, and trauma experienced by those in high-risk professions such as the military, police officers, firefighters, healthcare professionals and war correspondents. Where the trauma is chronic, it can result in **complex post-traumatic stress disorder (CPTSD)**, often causing shame, dissociation and emotional instability.

Prefrontal cortex

A section of the **cerebral cortex*** that covers the front part of the brain, behind the forehead. It is particularly large in humans, allowing us to think about the consequences of a range of possible future actions. Once we have made a decision, it also helps us organise ourselves to act on it. In people with chronic pain, the prefrontal cortex loses its ability to regulate fear and worry, leading to mental health problems such as **anxiety,*** **depression*** and suicidal tendencies. One section of the prefrontal cortex, **Brodmann area 47**, helps us anticipate what's going to happen, and is very active when we listen to music, trying to predict what the melody is going to do next.

Prosopagnosia

A condition experienced by 1 in 50 people in which the brain's ability to recognise human faces breaks down. If the prosopagnosia is severe, they can even have trouble recognising members of their own family. Prosopagnosia can result from a brain injury or neurodegenerative disease, or can be passed down within families.

Psilocybin

A psychedelic that is the active ingredient in 'magic mushrooms', used in the treatment of **depression.*** It acts in the same way as **LSD (lysergic acid diethylamide).***

Psychosis

A severe stress-related illness in which the person can lose contact with reality.

Psychosomatic illness

A disabling illness that cannot be explained by medical tests or physical examination, often not related to any other mental health problems. It differs from **hypochondria**, which is disabling anxiety about illness. People are disabled by the anxiety, as opposed to the symptoms they have. Scans have shown differences in the brains of people who have psychosomatic illness from the brains of people who are well, or people who are faking illness.

Recovery model

An approach to mental health therapy that takes people's personal life experience into account. It encourages them not to be held back by their mental health issues and to make decisions about their own healthcare. Clinical support moves from being at the centre to being just one resource among many, including support from the community.

Rehearsal

A process whereby things we do during the day that are important to us or hold some emotion are spontaneously replayed the next time we sleep. This strengthens the connections between the neurons involved in those daytime activities.

Schizophrenia

A mental disorder characterised by disruptions in thought processes, perceptions, emotional responsiveness and social interactions. It can involve difficulty in distinguishing between reality and hallucinations. Typically persistent, it can be both severe and disabling. Treatment often involves a combination of medications, psychotherapy and specialised care services.

Social dreaming

A movement in which people participate in a **social dreaming matrix**, either in person or online, sharing their dreams and exploring how one dream connects with others, and what that says about the broader social consciousness.

Structural plasticity

Changes in the brain brought about through learning and experience.

Synaesthesia

A range of different conditions that involve a 'mixing of the senses'. Common types are **grapheme colour synaesthesia**, in which letters and numbers evoke colours, and **auditory visual synaesthesia**, where different types of sounds create specific visual experiences. A third type is **mirror-touch synaesthesia**, in which the person experiences the pain and emotions of other people as if they're happening within the person's own body. It's believed to be an extreme form of empathy, which most humans feel.

Transcranial direct current stimulation (tDCS)
A non-invasive brain stimulation technique that uses a mild electrical current applied directly to the scalp to treat depression and other psychiatric disorders.

Transcranial magnetic stimulation (TMS)
A non-invasive brain stimulation technique using electromagnetic induction to treat depression and other psychiatric disorders.

Trauma
The response to an overwhelming event or events. These could include natural disasters, serious accidents, physical or sexual abuse or neglect. Trauma from a single event is referred to as **single-incident trauma**. An experience of many repeated traumatic events over time is called **complex post-traumatic stress disorder (CPTSD)**. **Vicarious trauma** happens when someone is exposed to another person's trauma.

Voice-hearing (auditory hallucinations)
The most common form of hallucination, often experienced by those with **schizophrenia*** or **psychosis.*** Recent research shows that they can also happen in conjunction with other mental disorders such as **depression*** and **anxiety,*** and even in otherwise mentally healthy people. In one study of a group of 30,000 people, about 1 in 20 reported having at least one hallucination across their lifetime.

REFERENCES

Introduction

'Lynne's Lungs and other Adventures', www.abc.net.au/radionational/
  programs/offtrack/lynnes-lungs-and-other-abc-adventures/12868642

Part I: New Insights into How the Brain and Mind Work

1. Our Changeable Brain

Books

Doidge, Norman, *The Brain That Changes Itself: Stories of Personal Triumph
  from the Frontiers of Brain Science*, Viking, 2007

Doidge, Norman, *The Brain's Way of Healing: Remarkable Discoveries and
  Recoveries from the Frontiers of Neuroplasticity,* Scribe, 2015

Elliott, Clark, *The Ghost in My Brain: How a Concussion Stole My Life and
  How the New Science of Brain Plasticity Helped Me Get It Back*, Viking,
  2015

Hebb, Donald O, *Organization of Behavior: A Neuropsychological Theory*,
  Wiley & Sons, 1949 (republished by Psychology Press, 2002)

Merzenich, Michael, *Soft-Wired: How the New Science of Brain Plasticity Can
  Change Your Life*, Parnassus, 2013

Articles

Elliot, Clark, 'The ghost in my brain: How I found myself again,
  eight years after concussion', *ABC Radio National*, 6 May 2016, www.
  abc.net.au/radionational/programs/allinthemind/clark-elliot-the-
  ghost-in-my-brain/7391502

Elwood, Peter, *et al*, 'Healthy lifestyles reduce the incidence of chronic
  diseases and dementia: Evidence from the Caerphilly Cohort Study',

PLoS ONE, Vol 8, No 12 (December 2013), e81877, doi.org/10.1371/journal.pone.0081877

Malcolm, Lynne, 'Can brain training really rewire your brain?', 27 November 2013, *ABC Radio National*, www.abc.net.au/radionational/programs/allinthemind/can-brain-training-really-rewire-your-brain/5120178

Malcolm, Lynne, 'Neuroplasticity: How the brain can heal itself', 21 April 2015, www.abc.net.au/radionational/programs/allinthemind/neuroplasticity-and-how-the-brain-can-heal-itself/6406736

Maxfield, Maree, Cooper, M.S., and Kavanagh, A. *et al*, 'On the outside looking in: A phenomenological study of the lived experience of Australian adults with a disorder of the corpus callosum',*Orphanet Journal of Rare Diseases*, Vol 16, Art No 512 (December 2021), doi.org/10.1186/s13023-021-02140-5

Websites

Australian Disorders of the Corpus Callosum (ausDoCC), ausdocc.org.au

Mind-Eye Institute, mindeye.com

Queensland Brain Institute, The University of Queensland Australia, qbi.uq.edu.au

Lynne's All in the Mind *episodes*

'Brain work', broadcast 24 November 2013, www.abc.net.au/radionational/programs/allinthemind/brain-work/5101276

'The ghost in my brain', broadcast 1 May 2016, www.abc.net.au/radionational/programs/allinthemind/the-ghost-in-my-brain/7360674

'The healing brain', broadcast 19 April 2015, abc.net.au/radionational/programs/allinthemind/the-healing-brain/6391744

'The mysterious corpus callosum', broadcast 8 May 2016, www.abc.net.au/radionational/programs/allinthemind/the-mysterious-corpus-callosum/7377346

2. The Mystery of Consciousness

Books

Chalmers, David J, *The Conscious Mind: In Search of a Fundamental Theory*, Oxford University Press, 1996

Eagleman, David, *The Brain: The Story of You*, Canongate, 2015

Goldstein, E Bruce, *The Mind: Consciousness, Prediction, and the Brain*, MIT Press, 2020

Pistorius, Martin, *Ghost Boy: My Escape from a Life Locked Inside My Own Body*, Simon & Schuster, 2011

Articles

Chalmers, David J, 'Facing up to the problem of consciousness', *Journal of Consciousness Studies*, Vol 2, No 3 (1995), pp 200–219

Owen, Adrian M, *et al*, 'Detecting awareness in the vegetative state', *Science*, Vol 313, No 5792 (September 2006), p 1402

Websites

Ghost Boy, www.martinpistorius.com

Lynne's All in the Mind *episodes*

'Locked in', broadcast May 2015, www.abc.net.au/radionational/programs/allinthemind/locked-in/12910518

'The predictive mind', broadcast 11 October 2020, www.abc.net.au/radionational/programs/allinthemind/the-predictive-mind/12740654

'The story of your brain', broadcast 31 January 2016, www.abc.net.au/radionational/programs/allinthemind/the-story-of-your-brain/7108384

3. Memory

Books

Corkin, Suzanne, *Permanent Present Tense: The Man with No Memory, and What He Taught the World*, Penguin, 2013

Kelly, Lynne, *The Memory Code: Unlocking the Secrets of the Lives of the Ancients and the Power of the Human Mind*, Atlantic Books, 2017

Ogden, Jenni, *Trouble in Mind: Stories from a Neuropsychologist's Casebook*, Oxford University Press, 2012

Reese, Elaine, *Tell Me a Story: Sharing Stories to Enrich Your Child's World*, Oxford University Press, 2013

Sharrock, Rebecca, *Streams of Memories: A Firsthand Account of What It Is Like to Remember Every Day from Infancy*, self-published, 2017

Articles

Layt, Stuart, 'New UQ clinic to help with early diagnosis of dementia', *Brisbane Times*, 21 September 2021, www.brisbanetimes.com.au/national/queensland/new-uq-clinic-to-help-with-early-diagnosis-of-dementia-20210921-p58tij.html

Marshall, Sean, and Reese, Elaine. 'Growing memories: Benefits of an early childhood maternal reminiscing intervention for emerging adults' turning point narratives and well-being', *Journal of Research in Personality*, Vol 99, 104262 (August 2022), doi.org/10.1016/j.jrp.2022.104262

Mitchell, Claire, and Reese, Elaine, 'Growing memories: Coaching mothers in elaborative reminiscing with toddlers benefits adolescents' turning point narratives and wellbeing', *Journal of Personality*, Vol 24, No 6 (January 2022), pp 887–901

Reese, Elaine, 'How you can talk to your toddler to safeguard their well-being when they grow into a teenager', *The Conversation*, 9 March, 2022, theconversation.com/how-you-can-talk-to-your-toddler-to-safeguard-their-well-being-when-they-grow-into-a-teenager-177536

Reese, Elaine, and Newcombe, Rhiannon, 'Training mothers in elaborative reminiscing enhances children's autobiographical memory and narrative', *Child Development*, Vol 78, No 4 (2007), pp 1153–1170. https://doi.org/10.1111/jopy.12703

Reese, Elaine, and Robertson, Sarah-Jane, 'Origins of adolescents' earliest memories', *Memory*, Vol 27, No 1 (2019), pp 79–91, doi-org.ezproxy.otago.ac.nz/10.1080/09658211.2018.1512631

Websites
Growing Up in New Zealand, www.growingup.co.nz

Lynne's All in the Mind *episodes*
'A highly superior memory', broadcast 11 February 2018, www.abc.net.au/radionational/programs/allinthemind/a-highly-superior-memory/9397666

'A history of memory', broadcast 18 November 2012, www.abc.net.au/radionational/programs/allinthemind/a-history-of-memory/4373748

'HM – The man with no memory', broadcast 10 November 2013, www.abc.net.au/radionational/programs/allinthemind/hm---the-man-with-no-memory/5067570

'The Indigenous memory code', broadcast 3 July 2016, www.abc.net.au/radionational/programs/allinthemind/indigenous-memory-code/7553976

'Memories and fears panel discussion from *Big Ideas* [radio program]', recorded at the World Science Festival, State Library of Queensland, 24 March 2018, and broadcast 14 May 2018, www.abc.net.au/radionational/programs/allinthemind/bia-memories-and-fears/9759488

'Memory: The thread of life', broadcast 16 December 2012, www.abc.net.au/radionational/programs/allinthemind/memory---the-thread-of-life/4409988

'The power of one', broadcast 3 November 2013, www.abc.net.au/radionational/programs/allinthemind/the-power-of-one/5053668

'The predictive mind', broadcast 11 October 2020, www.abc.net.au/radionational/programs/allinthemind/the-predictive-mind/12740654

'Remembering together', broadcast 27 May 2012, www.abc.net.au/
radionational/programs/allinthemind/new-document/4028890

4. Creativity

Books

Andreasen, Nancy C, *The Creating Brain: The Neuroscience of Genius*, Dana
Press, 2005

Eagleman, David, and Brandt, Anthony, *The Runaway Species: How
Human Creativity Remakes the World*, Canongate, 2017

Kyaga, Simon, *Creativity and Mental Illness: The Mad Genius in Question*,
Palgrave Macmillan, 2014

Articles

Andrews, Kylie, 'The Aha! Challenge: Using brain teasers to understand
eureka moments', *ABC News*, 9 August 2019, www.abc.net.au/
news/science/2019-08-09/aha-challenge-measures-insight-aha-
moments/11396746

Branchini, E, *et al*, 'Can contraries prompt intuition in insight problem
solving?', *Frontiers in Psychology*, 26 December 2016, www.frontiersin.
org/articles/10.3389/fpsyg.2016.01962/full

Danek, Amory H, *et al*, 'Aha! experiences leave a mark: Facilitated recall
of insight solutions', *Psychological Research,* Vol 77 (September 2013),
pp 659–669, www.amorydanek.de/wp-content/uploads/2019/02/
Danek-2013-Aha-exp-leave-a-mark.pdf

Kizilirmak, JM, *et al*, 'Generation and the subjective feeling of "aha!"
are independently related to learning from insight', *Psychological
Research*, Vol 80, No 6 (2016), pp 1059–1074, doi.org/10.1007/
s00426-015-0697-2

Malcolm, Lynne, and Andrews, Kylie, 'What drives creativity?
A complex network that scientists are striving to better
understand', *ABC News*, 17 August 2019, www.abc.net.au/news/
science/2019-08-17/creativity-neuroscience-your-brain/11420898

Martin, Brittany Harker, and Colp, S Mitchell, 'Art making promotes
mental health: A solution for schools that time forgot', *Canadian
Journal of Education*, Vol 45, No 1 (Spring 2022), pp 156–163

Olson, Jay A, *et al*, 'Naming unrelated words predicts creativity', *PNAS,*
Vol 118, No 25 (June 2021), www.pnas.org/doi/full/10.1073/
pnas.2022340118

Schou, Mogens, 'Artistic productivity and lithium prophylaxis in
manic-depressive illness', *British Journal of Psychiatry*, Vol 135 (1979),
pp 97–103

Shen, W *et al,* 'Feeling the insight: Uncovering Somatic Marjers of the "aha" Experience', *Applied Psychophysiology and Biofeedback,* Vol 43, pp 13–21 (2018), https://link.springer.com/article/10.1007/s10484-017-9381-1

Websites

Cognitive Neuroscience of Creativity Laboratory (CNCL), beatylab.la.psu.edu

David Eagleman: Neuroscientist, Author, Technologist, Entrepreneur, eagleman.com

The Divergent Association Task, www.datcreativity.com/

Lynne's All in the Mind *episodes*

'The creating brain – Reaching Xanadu', 4 March 2006, abc.net.au/ radionational/programs/allinthemind/the-creating-brain---reaching-xanadu/3301432

'Art, science and schizophrenia', broadcast 4 November 2012, www. abc.net.au/radionational/programs/allinthemind/art2c-science-and-schizophrenia/4340806

'Creativity and mental illness', broadcast 7 August 2016, www.abc.net. au/radionational/programs/allinthemind/creativity-and-mental-illness/7291448

'Creativity and the A-ha moment', broadcast 18 August 2019, www.abc. net.au/radionational/programs/allinthemind/creativity-and-the-a-ha-moment/11413166

'Creativity and your brain', broadcast 20 May 2018, www.abc.net. au/radionational/programs/allinthemind/creativity-and-your-brain/9758846

'Prescribing art for mental health', broadcast 9 August 2020, www.abc. net.au/radionational/programs/allinthemind/prescribing-art-for-mental-health/12523324

5. Sleep

Books

Leschziner, Guy, *The Nocturnal Brain: Nightmares, Neuroscience and the Secret World of Sleep,* Simon & Schuster, 2019

Lewis, Penelope A, *The Secret World of Sleep: The Surprising Science of the Mind at Rest,* St Martin's Press, 2014

Walker, Matthew, *Why We Sleep: Unlocking the Power of Sleep and Dreams,* Scribner, 2017

Articles

Boyce, Richard, *et al,* 'Causal evidence for the role of REM sleep theta rhythm in contextual memory consolidation', *Science,* Vol 352, No 6287 (13 May 2016), pp 812–816

Everson, CA, Bergmann, BM, and Rechtschaffen, A, 'Sleep deprivation in the rat: III. Total sleep deprivation', *Sleep*, Vol 12, No 1 (February 1989), pp 13–21

Lewis, Penelope A, Knoblich, Günther, and Poe, Gina, 'How memory replay in sleep boosts creative problem-solving', *Trends in Cognitive Sciences,* Vol 22, No 6 (June 2018), pp 491–503

Tempesta, Daniela, *et al,* 'Sleep and emotional processing', *Sleep Medicine Reviews*, Vol 40 (August 2018), pp 183–195

Van der Helm, Els, and Walker, Matthew P, 'Overnight therapy? The role of sleep in emotional processing', *Psychological Bulletin*, Vol 135, No 5 (September 2009), pp 731–748

Websites
The Williams Hippocampus and Memory Laboratory, Douglas Institute – McGill University, sylvainwilliams.ca

Lynne's All in the Mind *episodes*
'Adventures in sleep', broadcast 16 June 2019, www.abc.net.au/radionational/programs/allinthemind/adventures-in-sleep/11200022

'Dream sleep', broadcast 18 September 2016, www.abc.net.au/radionational/programs/allinthemind/dream-sleep/7840558

'The mind at rest', broadcast 15 December 2013, www.abc.net.au/radionational/programs/allinthemind/the-mind-at-rest/5141356

6. Dreams
Books
Freud, Sigmund, *Die Traumdeutung (The Interpretation of Dreams)*, Franz Deuticke, 1900

Grotstein, James S, *Who Is the Dreamer, Who Dreams the Dream? A Study of Psychic Presences*, Routledge, 2000

LaBerge, Stephen, *Exploring the World of Lucid Dreaming*, Ballantine, 1990

LaBerge, Stephen, *Lucid Dreaming: The Power of Being Awake and Aware in Your Dreams*, St Martin's Press, 1985

Long, Susan, and Manley, Julian (eds), *Social Dreaming: Philosophy, Research, Theory and Practice*, Routledge, 2019

Pagel, JF, *Dream Science: Exploring the Forms of Consciousness*, Academic Press, 2014

Zadra, Antonio, and Stickgold, Robert, *When Brains Dream: Exploring the Science and Mystery of Sleep*, Norton, 2021

Articles
Irish, M, Piguet, O, and Hodges, JR, 'Self-projection and the default

network in frontotemporal dementia', *Nature Reviews Neurology*, Vol 8, (2012), pp 152–161

O'Callaghan, C, *et al*, 'Hippocampal atrophy and intrinsic brain network dysfunction relate to alterations in mind wandering in neurodegeneration', *Proceedings of the National Academy of Sciences*, Vol 116, No 8 (2019), pp 3316–3321

Pagel, James F, and Kwiatkowski, CF, 'Creativity and dreaming: Correlation of reported dream incorporation into waking behavior with level and type of creative interest', *Creativity Research Journal*, Vol 15, Nos 2–3 (July 2003), pp 199–205

Websites

International Association for the Study of Dreams (ASD International), www. asdreams.org

Social Dreaming International Network, socialdreaminginternational.net

Lynne's **All in the Mind** *episodes*

'The bizarre dreaming of COVID-19', broadcast 30 August 2020, www.abc.net.au/radionational/programs/allinthemind/the-bizarre-dreaming-of-covid-19/12594016

'Dementia, sleep and daydreaming', broadcast 28 April 2019, www.abc.net.au/radionational/programs/allinthemind/dementia-sleep-and-daydreaming/11038862

'Dream sleep', broadcast 18 September 2016, www.abc.net.au/radionational/programs/allinthemind/dream-sleep/7840558

'Dreams – The lucid experience', broadcast 2 November 2014, www.abc.net.au/radionational/programs/allinthemind/5844134

'Dreams – Windows to the mind', broadcast 18 January 2015, www.abc.net.au/radionational/programs/allinthemind/dreams-pt1/5967256

'The mind at rest', broadcast 15 December 2013, www.abc.net.au/radionational/programs/allinthemind/the-mind-at-rest/5141356

'Sharing dreams and social visions', broadcast 6 September 2020, www.abc.net.au/radionational/programs/allinthemind/social-dreaming/12607932

Part II: New Insights into Living with Mental 'Difference'

7. Autism and 'Neurodiversity'

Books

Belcher, Dr Hannah Louise, *Taking Off the Mask: Practical Exercises to Help Understand and Minimise the Effects of Autistic Camouflaging*, Jessica Kingsley Publishers, 2022

Bennett, Jill (ed), *The Big Anxiety: Taking Care of Mental Health in Times of Crisis*, Bloomsbury, 2022

Carpenter, Barry, Happé, Francesca, and Egerton, Jo (eds), *Girls and Autism: Educational, Family and Personal Perspectives*, Routledge, 2019

Purkis, Jeanette, Goodall, Dr Emma, and Nugent, Dr Jane, *The Guide to Good Mental Health on the Autism Spectrum*, Jessica Kingsley Publishers, 2016

Robison, John Elder, *Switched On: A Memoir of Brain Change and Emotional Awakening*, Spiegel & Grau, 2016

Silberman, Steve, *NeuroTribes: The Legacy of Autism and the Future of Neurodiversity*, Avery, 2015

Articles

Amaral, David G, 'Language in *Autism Research*: Accurate and respectful', Wiley Online Library, 30 December 2022, https://hubs.li/Q01x5nMj0

Happé, F and Baron-Cohen, S, 'Remembering Lorna Wing (1928–2014)', *Spectrum*, 15 July 2014, www.spectrumnews.org/opinion/remembering-lorna-wing-1928-2014/

Malcolm, Lynne, and Willis, Olivia, 'The journey to understanding autism', *ABC Radio National*, 3 November 2015, www.abc.net.au/radionational/programs/allinthemind/the-journey-to-understanding-autism/6908040

Malcolm, Lynne, and Willis, Olivia, 'Why autism spectrum disorders are under-diagnosed in women and girls', *ABC Radio National*, 25 June 2015, www.abc.net.au/radionational/programs/allinthemind/why-autism-spectrum-disorders-are-under-diagnosed-in-women/6570896

Wing, L, and Gould, J, 'Severe impairments of social interaction and associated abnormalities in children: epidemiology and classification', *Journal of Autism and Developmental Disorders*, Vol 9, No 1 (March 1979), pp 11–29, pubmed.ncbi.nlm.nih.gov/155684

Websites

A_tistic: Theatre – Workshops – Consultation, www.a-tistic.com.au

Autism Spectrum Australia, www.autismspectrum.org.au

The Big Anxiety: An Initiative of UNSW, thebiganxiety.org, Big Anxiety Research Centre

Laser Beak Man by Tim Sharp, laserbeakman.com

Look Me in the Eye: Official Blog of NYT Bestselling Author, Photographer, Educator, Neurodiversity Advocate and Automobile Aficionado John Elder Robison, jerobison.blogspot.com

Lynne's All in the Mind *episodes*

'Apps for autism', broadcast 14 August 2016, www.abc.net.au/radionational/programs/allinthemind/apps-for-autism/7701834

'The art of neurodiversity', broadcast March 2018, www.abc.
net.au/radionational/programs/allinthemind/the-art-of-neurodiversity/10571842

'Autism and empowerment', broadcast 12 June 2016, www.abc.net.au/
radionational/programs/allinthemind/autism/7485264

'Autism, horses and other therapies', broadcast 19 October 2014, www.
abc.net.au/radionational/programs/allinthemind/autism2c-horses-and-other-therapies/5809256

'Autism and superheroes', broadcast 29 September 2019, www.
abc.net.au/radionational/programs/allinthemind/autism-and-superheroes/11546168

'Girls and autism', broadcast 21 June 2015, www.abc.net.au/
radionational/programs/allinthemind/girls-and-autism/6549090

'Neurotribes', broadcast 1 November 2015, www.abc.net.au/
radionational/programs/allinthemind/neurotribes/6887954

'On the spectrum', broadcast 5 May 2013, www.abc.net.au/radionational/
programs/allinthemind/on-the-spectrum/4652550

'The social brain', broadcast 28 April 2013, www.abc.net.au/
radionational/programs/allinthemind/the-social-brain/4646808

'Tuning in to autism', broadcast 11 September 2016, www.abc.net.au/
radionational/programs/allinthemind/tuning-in-to-autism/7819618

8. Hearing Colours, Feeling Other People's Pain

Books

Thomson, Helen, *Unthinkable: An Extraordinary Journey Through the World's Strangest Brains*, Ecco, 2018

Articles

Berger, Joshua J, *et al*, 'Sharing the load: How a personally coloured calculator for grapheme-colour synaesthetes can reduce processing costs', *PLoS ONE*, Vol 16, No 9 (September 2021), e0257713, doi.org/10.1371/journal.pone.0257713

Maguire, Sarah, and Rich, Anina (researcher), '"Mind-reading" test validates remarkable world of synaesthetes', 24 February 2021, *The Lighthouse* (Macquarie University), lighthouse.mq.edu.au/article/february-2021/Mind-reading-test-validates-remarkable-world-of-synaesthetes

Teichmann, Lina, *et al*, 'Temporal dissociation of neural activity underlying synesthetic and perceptual colors', *PNAS*, Vol 118, No 6 (February 2021), e2020434118, doi.org/10.1073/pnas.2020434118

Videos

Power Institute, 'The taste of purple: An evening of synaesthesia', 17 May 2018, *YouTube*, www.youtube.com/watch?v=oiDo6PKhR8E

Websites

Anina Rich, Macquarie University Anina Rich, researchers.mq.edu.au/en/persons/anina-rich

Ninanovart, www.ninanovart.com

Oliver Sacks Foundation, www.oliversacks.com

Lynne's **All in the Mind** *episodes*

'Strange brains and rare perceptions', broadcast 3 June 2018, www.abc.net.au/radionational/programs/allinthemind/unthinkable/9812624

'Synesthesia and art', broadcast 17 June 2018, www.abc.net.au/radionational/programs/allinthemind/synesthesia-and-art/9860908

'Synesthesia: seeing sounds, hearing colours', broadcast 10 June 2018, www.abc.net.au/radionational/programs/allinthemind/synesthesia:-seeing-sounds-and-hearing-colours/9836826

9. Faced With Strangers

Books

Sacks, Oliver, *The Man Who Mistook His Wife for a Hat and Other Clinical Tales*, Duckworth, 1985

Sacks, Oliver, *The Mind's Eye*, Knopf, 2010

Articles

Dunn, James D, *et al*, 'UNSW Face Test: A screening tool for super-recognizers', *PLoS ONE*, Vol 15, No 11 (November 2020), e0241747, doi.org/10.1371/journal.pone.0241747

Gomez, Jesse, *et al*, 'Microstructural proliferation in human cortex is coupled with the development of face processing', *Science*, Vol 355, No 6320 (January 2017), pp 68–71

Russell, Richard, Duchaine, Brad, and Nakayama, Ken, 'Super-recognizers: People with extraordinary face recognition ability', *Psychonomic Bulletin & Review*, Vol 16 (2009), pp 252–257, doi.org/10.3758/PBR.16.2.252

Sacks, Oliver, 'Face-blind: Why are some of us terrible at recognizing faces?', *The New Yorker*, 23 August 2010, www.newyorker.com/magazine/2010/08/30/face-blind

Schmalzl, Laura, *et al*, 'Training of familiar face recognition and visual scan paths for faces in a child with congenital prosopagnosia', *Cognitive Neuropsychology*, Vol 25, No 5 (July 2008), pp 704–729

Videos

'Dr Karl on face blindness', ABC Science, 16 February 2017, *YouTube*,
www.youtube.com/watch?v=FWKfd0qOgUs

Websites

UNSW Face Test, facetest.psy.unsw.edu.au

Lynne's All in the Mind *episodes*

'Super-recognisers', broadcast 11 March 2018, www.abc.net.au/
radionational/programs/allinthemind/super-recognisers/9523296

'What's in a face?', broadcast 18 February 2006, www.abc.net.au/
radionational/programs/allinthemind/whats-in-a-face/3309304

'What's in a face? Prosopagnosia', broadcast 19 February 2017, www.
abc.net.au/radionational/programs/allinthemind/whats-in-a-face-
prosopagnosia/8269742

10. Seeing When You're Blind

Websites

Charles Bonnet Syndrome Foundation, www.charlesbonnetsyndrome.org

Videos

Sacks, Oliver, 'What hallucinations reveal about our minds', *TED*, www.
ted.com/talks/oliver_sacks_what_hallucination_reveals_about_our_
minds

Lynne's All in the Mind *episodes*

'Seeing when you're blind', broadcast 2 August 2020, www.abc.net.
au/radionational/programs/allinthemind/seeing-when-youre-
blind/12503398

Part III: New Insights into Debilitating Mental Conditions

11. Dementia

Books

Bryden, Christine, *Before I Forget: How I Survived Being Diagnosed with
Younger-Onset Dementia at 46*, Penguin, 2015

Genova, Lisa, *Still Alice: A Novel*, Gallery Books, 2009

Montague, Jules, *Lost and Found: Memory, Identity and Who We Become
When We're No Longer Ourselves*, Hodder & Stoughton, 2018

Roberts, Dr Kailas, *Mind Your Brain: The Essential Australian Guide to
Dementia*, University of Queensland Press, 2021

Articles

Irish, M, 'The self in dementia is not lost, and can be reached with care',
Aeon, 13 November 2019

Strikwerda-Brown, C, *et al*, '"All is not lost" – Rethinking the nature of memory and the self in dementia', *Ageing Research Reviews*, Vol 54 (2019), 100932

Strohminger, Nina, and Nichols, Shaun, 'Neurodegeneration and identity', *Psychological Science*, Vol 26, No 9 (September 2015), pp 1469–1479

Websites
Australian Government: Australian Institute of Health and Welfare (AIHW), www.aihw.gov.au

Christine Bryden, christinebryden.com

Dementia Australia, www.dementia.org.au

Dementia Support Australia (DSA), www.dementia.com.au

'Memory and age', *Queensland Brain Institute, The University of Queensland Australia*, qbi.uq.edu.au/brain-basics/memory/memory-and-age

Your Brain in Mind, www.yourbraininmind.com.au

Lynne's All in the Mind *episodes*
'Before I forget', broadcast 13 December 2015, www.abc.net.au/radionational/programs/allinthemind/7012970

'Dementia, sleep and daydreaming', broadcast 28 April 2019, www.abc.net.au/radionational/programs/allinthemind/dementia-sleep-and-daydreaming/11038862

'The funny side of Alzheimer's', broadcast 5 June 2016, www.abc.net.au/radionational/programs/allinthemind/the-funny-side-of-alzheimers/7463094

'Memory loss and identity', broadcast 19 August 2018, www.abc.net.au/radionational/programs/allinthemind/memory-loss-and-identity/10119238

'*Still Alice* and other stories of neuroscience', broadcast 28 February 2016, www.abc.net.au/radionational/programs/allinthemind/still-alice-and-other-stories/7197028

12. Anxiety

Books
Lowinger, Jodie, *The Mind Strength Method: Four steps to curb anxiety, conquer worry & build resilience,* Murdoch Books, 2021

Websites
Dr Jodie: Mind Strength Method, drjodie.com.au

SANE, www.sane.org

Lynne's All in the Mind *Episodes*

'Anxiety – And the "worry bully"', broadcast 15 September 2019, www. abc.net.au/radionational/programs/allinthemind/anxiety---and-the-worry-bully/11493198

'Facing fears and phobias', broadcast 24 March 2019, www.abc. net.au/radionational/programs/allinthemind/facing-fears-and-phobias/10916714

'The heritability of mental illness', broadcast 30 October 2016, www. abc.net.au/radionational/programs/allinthemind/the-heritability-of-mental-illness/7885506

'Monkey mind', broadcast 21 April 2013, www.abc.net.au/radionational/programs/allinthemind/monkey-mind/4617718

'Orchids, dandelions and an intriguing set of genes', broadcast 22 April 2012, www.abc.net.au/radionational/programs/allinthemind/orchids-dandelions-genes/3958630

'The other end of shyness', broadcast 2006, www.abc.net.au/radionational/programs/allinthemind/the-other-end-of-shyness/4083684

'Parenting with a mental illness', broadcast 20 December 2020, www. abc.net.au/radionational/programs/allinthemind/parenting-with-a-mental-illness/12943094

'Reflections on shame', broadcast 23 August 2020, www.abc.net.au/radionational/programs/allinthemind/shame/12574742

'The secret history of self-harm', broadcast 23 April 2017, www.abc.net. au/radionational/programs/allinthemind/a-secret-history-of-self-harm/8453616

'Transitioning to motherhood: Perinatal mental health', broadcast 11 November 2018, www.abc.net.au/radionational/programs/allinthemind/perinatal-mental-health/10478792

'Women's mental health and hormones', broadcast 22 November 2015, www.abc.net.au/radionational/programs/allinthemind/womens-mental-health/6945238

'Young minds, the highs and lows', broadcast 10 June 2012, www.abc. net.au/radionational/programs/allinthemind/young-minds2c-the-highs-and-lows/4054982

13. Depression

Books

Korb, Alex, PhD, *The Upward Spiral: Using Neuroscience to Reverse the Course of Depression, One Small Change at a Time*, New Harbinger, 2015

Articles

Malcolm, Lynne, and Willis, Olivia, 'A new era of depression treatments', *ABC Radio National*, 14 August 2015, www.abc.net.au/radionational/programs/allinthemind/a-new-era-of-depression-treatments/6697434

Sparkes, Michelle, 'Help for depression with Prof Ian Hickie AM' (article and video), *Michelle Sparkes*, michellesparkes.com/depression-w-prof-ian-hickie

Lynne's All in the Mind *episodes*

'Depression and #ReasonsToStayAlive', broadcast 31 May 2015, www.abc.net.au/radionational/programs/allinthemind/depression-and-reasons-to-stay-alive/6494972

'Depression and your sense of self', broadcast 29 July 2018, www.abc.net.au/radionational/programs/allinthemind/depression-and-the-self/10034452

'Depression treatment: The way forward', broadcast 9 August 2015, www.abc.net.au/radionational/programs/allinthemind/depression/6667952

'The genetics of depression', broadcast 25 June 2017, www.abc.net.au/radionational/programs/allinthemind/genetics-of-depression/8635228

'How to stay mentally healthy', broadcast 27 September 2020, www.abc.net.au/radionational/programs/allinthemind/staying-mentally-healthy/12688254

'Podcast extra: Dr Alex Korb offers more techniques out of depression, anxiety', broadcast 28 September 2020, www.abc.net.au/radionational/programs/allinthemind/alex-korb-pod-interview/12688654

'Ruby's sane world' (video), broadcast 26 May 2013, www.abc.net.au/radionational/programs/allinthemind/ruby27s-sane-world/4703592

'The scientist, the monk and Ruby Wax', broadcast 4 February 2018, www.abc.net.au/radionational/programs/allinthemind/the-scientist,-the-monk,-and-ruby-wax/9375802

'Taming the black dog – New approaches to depression', broadcast 7 December 2014, www.abc.net.au/radionational/programs/allinthemind/depression-new-therapies/5930868

14. Post-Traumatic Stress Disorder (PTSD)

Books

Herman, Judith, *Trauma and Recovery: The Aftermath of Violence: From Domestic Abuse to Political Terror*, Basic Books, 1992

Nelson, Charles A, Fox, Nathan A, and Zeanah, Charles H, *Romania's Abandoned Children: Deprivation, Brain Development, and the Struggle for Recovery*, Harvard University Press, 2014

Ruckel, Izidor, and Padbury, Sarah (ed.), *Izidor: The autobiography of Izidor Ruckel*, Ruckel International, April 2023

Van der Kolk, Bessel, MD, *The Body Keeps the Score: Brain, Mind, and Body in the Healing of Trauma*, Viking, 2014

Articles

Malcolm, Lynne, 'Healing trauma through mind and body', *ABC Radio National*, 29 April 2015, www.abc.net.au/radionational/programs/allinthemind/healing-trauma-through-the-mind-and-the-body/6425638

Malcolm, Lynne, 'When trauma becomes post-traumatic stress disorder', *ABC Radio National*, 28 April 2014, www.abc.net.au/radionational/programs/allinthemind/5415206

Websites

Bucharest Early Intervention Project, www.bucharestearlyinterventionproject.org

Center on the Developing Child – Harvard University, developingchild.harvard.edu

Izidor Ruckel: Orphan Advocate, izidorruckel.com

Police Post Trauma Support Group, www.pptsg.org.au

Trauma Centre Australia, traumacentre.com.au

Lynne's All in the Mind *episodes*

'Romania's abandoned children', broadcast 6 July 2014, www.abc.net.au/radionational/programs/allinthemind/romania27s-abandoned-children/5502878

'The scene of the crime: One too many? PTSD Part 2', broadcast 16 October 2004, www.abc.net.au/radionational/programs/allinthemind/the-scene-of-the-crime-one-too-many-ptsd-part-2/3426348

'Too much reality', broadcast 27 April 2014, www.abc.net.au/radionational/programs/allinthemind/too-much-reality/5385036

'Trauma, memory, and mental health', broadcast 18 March 2018, www.abc.net.au/radionational/programs/allinthemind/trauma,-memory,-and-health/9547446

'Trauma treatment', broadcast 26 April 2015, www.abc.net.au/radionational/programs/allinthemind/trauma-treatment/6405674

'When trauma tips you over: PTSD Part 1', broadcast 9 October 2004,
www.abc.net.au/radionational/programs/allinthemind/when-trauma-
tips-you-over-ptsd-part-1/3426554

15. Borderline Personality Disorder (BPD)

Books

Beatson, J, and Rao, S (eds), *Borderline Personality Disorder: A Practical
Guide for General Practitioners*, Spectrum Personality Disorder Service,
Richmond, Victoria, 2020

Articles

Paris, Joel, 'Suicidality in borderline personality disorder', *Medicina*
(Kaunas, Lithuania), Vol 55, No 6 (June 2019), p 223, www.ncbi.nlm.
nih.gov/pmc/articles/PMC6632023

Websites

Australian BPD Foundation Limited, www.bpdfoundation.org.au

Spectrum: Specialising in Personality Disorder and Complex Trauma, www.
spectrumbpd.com.au

Lynne's **All in the Mind** *Episodes*

'Borderline personality disorder – What works?', broadcast 23 November
2014, www.abc.net.au/radionational/programs/allinthemind/
borderline-personality-disorder/5881922

'I hurt myself: The secrecy of self harm', broadcast 29 April 2012, www.
abc.net.au/radionational/programs/allinthemind/all-in-the-mind-
29.4.2012-i-hurt-myself/3968842

'Mothering and mental illness', broadcast 13 January 2019, www.abc.net.
au/radionational/programs/allinthemind/mothering-and-mental-
illness/10571926

'Over the borderline' (transcript only), broadcast 20 November 2004,
www.abc.net.au/radionational/programs/allinthemind/over-the-
borderline/3427184

'A roller-coaster of emotion – Borderline personality disorder',
broadcast 6 October 2019, www.abc.net.au/radionational/programs/
allinthemind/a-roller-coaster-of-emotion-borderline-personality-
disorder/11564266

'Turbulent minds collide', broadcast 2 April 2017, www.abc.net.au/
radionational/programs/allinthemind/turbulent-minds-collide/8394188

16. Hearing Voices

Books

Fernyhough, Charles, *The Voices Within: The History and Science of How
We Talk to Ourselves*, Basic Books, 2016

Articles

Copolov, David L, Mackinnon, Andrew, and Trauer, Tom, 'Correlates of the affective impact of auditory hallucinations in psychotic disorders', *Schizophrenia Bulletin*, Vol 30, No 1 (2004), pp 163–171

Daya, Indigo, 'Russian dolls and epistemic crypts: A lived experience reflection on epistemic injustice and psychiatric confinement', *Incarceration* Vol 3, No 2 (July 2022), https://journals.sagepub.com/doi/10.1177/26326663221103445

Goozee, Rhianna, 'Hearing Voices: Tracing the borders of normality', *The Lancet Psychiatry*, Vol 2, No 3 (March 2015) pp 206–207, www.thelancet.com/journals/lanpsy/article/PIIS2215-0366(15)00066-8/fulltext

McGrath, John J, *et al*, 'Psychotic experiences in the general population: A cross-national analysis based on 31,261 respondents from 18 countries', *JAMA Psychiatry*, Vol 72, No 7 (July 2015), pp 697–705

Silver, Katie, 'Why it's healthy to give imaginary voices their own Facebook pages', *ABC Radio National*, 3 December 2013, www.abc.net.au/radionational/programs/allinthemind/5125556

Videos

Longden, Eleanor, 'The voices in my head', *TED*, www.ted.com/talks/eleanor_longden_the_voices_in_my_head

Websites

Hearing the Voice: Interdisciplinary Voice-Hearing Research, hearingthevoice.org

Hearing Voices Network: For People Who Hear Voices, See Visions or Have Other Unusual Perceptions, www.hearing-voices.org

The International Society for Psychological and Social Approaches to Psychosis (ISPS), www.isps.org

The International Society for Psychological and Social Approaches to Psychosis (ISPS) Australia, ispsaustralia.com

Intervoice: Connecting People and Ideas in the Hearing Voices Movement, www.intervoiceonline.org

Lynne's All in the Mind *episodes*

'Children who hear voices', broadcast 26 March 2017, www.abc.net.au/radionational/programs/allinthemind/children-who-hear-voices/8377416

'Compassion therapy for voice-hearing', broadcast 29 April 2018, www.abc.net.au/radionational/programs/allinthemind/compassion-therapy-for-voice-hearing/9692796

'Hearing voices – The invisible intruders', broadcast 22 July 2006, www.abc.net.au/radionational/programs/allinthemind/hearing-voices---the-invisible-intruders/3321982

'Invisible intruders – The voices in my head', broadcast 1 December 2013, www.abc.net.au/radionational/programs/allinthemind/invisible-intruders-e28093-the-voices-in-my-head/5117208

'Love, rock 'n' roll and the "S" word', broadcast 12 July 2015, www.abc.net.au/radionational/programs/allinthemind/love2c-rock-27n27-roll2c-and-the-27s27-word/6601136

'Our inner voices', broadcast 19 June 2016, www.abc.net.au/radionational/programs/allinthemind/our-inner-voices/7518138

'Schizophrenia: New clues', broadcast 5 October 2014, www.abc.net.au/radionational/programs/allinthemind/schizophrenia/5654952

17. Intergenerational Trauma

Articles

Interview with Tracy Westerman, *The Drum*, 9 September 2021

Menzies, Karen, 'Understanding the Australian Aboriginal experience of collective, historical and intergenerational trauma', *International Social Work*, Vol 62, No 6 (September 2019), pp 1522–1534, doi.org/10.1177/0020872819870585

Nasir, Bushra F *et al*, 'Traumatic life event and risk of post-traumatic stress disorder among the Indigenous population of regional, remote and metropolitan Central-Eastern Australia: a cross-sectional study', *BMJ Open*, Vol 11, No 4, https://bmjopen.bmj.com/content/11/4/e040875

Yehuda, Rachel, and Lehrner, Amy, 'Intergenerational transmission of trauma effects: Putative role of epigenetic mechanisms', *World Psychiatry*, Vol 17, No 3 (October 2018), pp 243–257

Websites

AIHW, 'Indigenous injury deaths, 2011–12 to 2015–16', 13 March 2020, www.aihw.gov.au/reports/injury/indigenous-injury-deaths-2011-12-to-2015-16/summary

Indigenous Psychological Services, https://indigenouspsychservices.com.au/

Lynne's All in the Mind *episodes*

'Indigenous hope and recovery', broadcast 5 July 2015, www.abc.net.au/radionational/programs/allinthemind/indigenous-hope-and-recovery/6583632

'Preventing Indigenous suicide', broadcast 8 November 2020, www.abc.net.au/radionational/programs/allinthemind/preventing-indigenous-suicide/12843508

'The strength of recognition', broadcast 19 March 2017, www.abc.
net.au/radionational/programs/allinthemind/the-strength-of-
recognition/8352904

18. Dissociative Identity Disorder (DID)
Books
Nathan, Debbie, *Sybil Exposed: The Extraordinary Story Behind the Famous Multiple Personality Case*, Simon and Schuster, 2011
Macken, Rhonda, *Hope Street: A Memoir of Multiple Personalities; Creating Selves to Survive*, Publicious, 2018

Articles
Kallena, 'What it's like to live with dissociative identity disorder', *ABC Radio National*, 1 March 2017, www.abc.net.au/news/2017-03-01/
what-its-like-to-live-with-dissociative-identity-disorder/8312076

Websites
Blue Knot Foundation : Empowering Recovery from Complex Trauma, blueknot.org.au

'Trauma and dissociation', *Belmont Private Hospital*, belmontprivate.com.
au/specialties/trauma-and-dissociation

'Dissociative Identity Disorder (DID)', *SANE*, www.sane.org/information-
and-resources/facts-and-guides/dissociative-identity-disorder

Lynne's All in the Mind *Episodes*
'Creating selves to survive', broadcast 27 October 2019, www.abc.
net.au/radionational/programs/allinthemind/creating-selves-to-
survive/11570864

'Dissociation and coping with trauma', broadcast 26 February 2017,
www.abc.net.au/radionational/programs/allinthemind/dissociation-
and-coping-with-trauma/8290294

'A superhuman escape', broadcast 16 July 2017, www.abc.net.au/
radionational/programs/allinthemind/a-superhuman-escape/8698400

Part IV: New Pathways to Healing
19. Revolutionary New Treatments for Mental Illness
Articles
Malcolm, Lynne, 'New approaches to treating depression',
ABC Radio National, 11 December 2014, www.abc.net.au/
radionational/programs/allinthemind/new-approaches-to-treating-
depression/5953642

Malcolm, Lynne, 'The treatment of trauma and the power of
neurofeedback', STARTTS Refugee Transitions newsletter, Issue 36,

February 2022, www.startts.org.au/media/The-treatment-of-trauma-and-the-power-of-Neurofeedback.pdf

Mitchell, Jennifer M, *et al*, 'MDMA-assisted therapy for severe PTSD: A randomized, double-blind, placebo-controlled phase 3 study', *Nature Medicine*, Vol 27 (May 2021), pp 1025–1033, doi.org/10.1038/s41591-021-01336-3

Polito, Vince, and Liknaitzky, Paul, 'The emerging science of microdosing: A systematic review of research on low dose psychedelics (1955–2021) and recommendations for the field', *Neuroscience & Biobehavioral Reviews*, Vol 139 (August 2022), 104706, doi.org/10.1016/j.neubiorev.2022.104706

Polito, Vince, and Stevenson, Richard J, 'A systematic study of microdosing psychedelics', *PLoS ONE*, Vol 14, No 2 (February 2019), e0211023, doi.org/10.1371/journal.pone.0211023

Tullis, Paul, 'How ecstasy and psilocybin are shaking up psychiatry', *Nature*, 27 January 2021, www.nature.com/articles/d41586-021-00187-9

Websites

'Daniel B. Fisher, M.D., Ph.D.', *National Empowerment Centre,* power2u.org/dan-fisher

'Psychedelics research and psilocybin therapy', *Psychiatry and Behavioral Sciences, John Hopkins Medicine*, www.hopkinsmedicine.org/psychiatry/research/psychedelics-research.html

TMS Clinics Australia, www.tmsaustralia.com.au

Lynne's All in the Mind *Episodes*

'Brain stimulation for depression', broadcast 27 December 2017, www.abc.net.au/radionational/programs/allinthemind/brain-stimulation-for-depression/9258614

'Brain training for the mentally ill', broadcast May 2014, www.abc.net.au/radionational/programs/allinthemind/brain-training/5433308

'Deep brain stimulation', broadcast 30 November 2014, www.abc.net.au/radionational/programs/allinthemind/deep-brain-stimulation/5909274

'Emotional CPR', broadcast 4 December 2016, www.abc.net.au/radionational/programs/allinthemind/emotional-cpr/8077910

'The food–mood connection', broadcast 14 May 2017, www.abc.net.au/radionational/programs/allinthemind/the-food-mood-connection/8510518

'In the therapy room', broadcast 30 April 2017, www.abc.net.au/radionational/programs/allinthemind/in-the-therapy-room/8197438

'Lived experience in mental health care', broadcast 5 November 2017, www.abc.net.au/radionational/programs/allinthemind/lived-experience-in-mental-health-care/9107916

'Machines for mental health', broadcast 23 October 2016, www.abc.net.au/radionational/programs/allinthemind/machines-for-mental-health/7885496

'Open dialogue', broadcast 21 February 2016, www.abc.net.au/radionational/programs/allinthemind/open-dialogue/7174084

'Optimism and hope – With Martin Seligman', broadcast 1 July 2018, www.abc.net.au/radionational/programs/allinthemind/optimism-and-hope---with-martin-seligman/9910458

'Positive psychology – With Martin Seligman', broadcast 24 June 2018, www.abc.net.au/radionational/programs/allinthemind/martin-seligman-1/9886008

'Psychedelics, addiction, and mental health', broadcast 24 February 2019, www.abc.net.au/radionational/programs/allinthemind/david-nutt/10829146

'The talking cure', broadcast 24 May 2015, www.abc.net.au/radionational/programs/allinthemind/the-talking-cure/6478848

'Therapies for OCD', broadcast 22 May 2016, www.abc.net.au/radionational/programs/allinthemind/therapies-for-ocd/7418052

'Too much reality', broadcast 27 April 2014, www.abc.net.au/radionational/programs/allinthemind/too-much-reality/5385036

'Tripping for depression', broadcast 4 August 2019, www.abc.net.au/radionational/programs/allinthemind/tripping-for-depression/11335926

'Tuning in to autism', 11 September 2016, www.abc.net.au/radionational/programs/allinthemind/tuning-in-to-autism/7819618

'Turn on, tune in', broadcast 28 July 2019, https://www.abc.net.au/radionational/programs/allinthemind/turn-on,-tune-in/11335924

20. Revolutionary New Treatments for Dementia
Books

Low, Lee-Fay, *Live and Laugh with Dementia: The Essential Guide to Maximizing Quality of Life*, Exisle, 2014

Low, Lee-Fay, and Laver, Kate (eds), *Dementia Rehabilitation: Evidence-Based Interventions and Clinical Recommendations*, Elsevier, 2020

Roberts, Dr Kailas, *Mind Your Brain: The Essential Australian Guide to Dementia*, University of Queensland Press, 2021

Sessa, B, *The Psychedelic Renaissance*, Aeon Books, 2012

Articles

Irish, M, (2022). 'Autobiographical memory in dementia syndromes –
an integrative review', *Wiley Interdisciplinary Reviews: Cognitive Science*,
14 October 2011, e1630, doi: 10.1002/wcs.1630

Low, Lee-Fay, *et al*, 'The Sydney Multisite Intervention of
LaughterBosses and ElderClowns (SMILE) study: Cluster randomised
trial of humour therapy in nursing homes', *BMJ Open*, Vol 3, No 1
(January 2013), pp 1–10, doi.org/10.1136/bmjopen-2012-002072

Strikwerda-Brown, C, et al, '"All is not lost" – Rethinking the nature
of memory and the self in dementia', *Ageing Research Reviews*, Vol 54
(2019), 100932

Strohminger, Nina, and Nichols, Shaun, 'Neurodegeneration and identity',
Psychological Science, Vol 26, No 9 (September 2015), pp 1469–1479

Voisey-Barlin, Maurie, Kennedy, Susan, and Low, Lee-Fay, 'Connecting
with Zoom Mates', *The Australian Journal of Dementia Care*, 28 October
2021, https://journalofdementiacare.com/connecting-with-zoom-
mates/

Websites

Australian Government: Australian Institute of Health and Welfare (AIHW),
www.aihw.gov.au

Dementia Australia, www.dementia.org.au

Dementia Support Australia (DSA), www.dementia.com.au

The Humour Foundation, www.humourfoundation.org.au

The Outside In Collective, www.facebook.com/outsideincollective/

'Memory and age', *Queensland Brain Institute, The University of Queensland
Australia*, qbi.uq.edu.au/brain-basics/memory/memory-and-age

Your Brain in Mind, www.yourbraininmind.com.au

Lynne's All in the Mind *episodes*

'Dementia care', broadcast 8 December 2013, www.abc.net.au/
radionational/programs/allinthemind/dementia-care/5125550

'Dementia, sleep and daydreaming', broadcast 28 April 2019, www.abc.
net.au/radionational/programs/allinthemind/dementia-sleep-and-
daydreaming/11038862

'The funny side of Alzheimer's', broadcast 5 June 2016, www.abc.
net.au/radionational/programs/allinthemind/the-funny-side-of-
alzheimers/7463094

'Memory loss and identity', broadcast 19 August 2018, www.abc.
net.au/radionational/programs/allinthemind/memory-loss-and-
identity/10119238

21. The Mind–Body Connection

Books

Benson, Herbert, MD, with Klipper, Miriam Z, *The Relaxation Response*, William Morrow, 1975

Harvey, Shannon, *The Whole Health Life: How You Can Learn to Get Healthy, Find Balance and Live Better in the Crazy, Busy, Modern World*, Whole Health Publishing, 2016

Hassed, Craig, and McKenzie, Stephen, *Mindfulness for Life*, Exisle Publishing, 2021

Jelinek, George, *Overcoming Multiple Sclerosis: The Evidence-based 7 Step Recovery Program*, Allen & Unwin, 2016

Marchant, Jo, *Cure: A Journey into the Science of Mind Over Body*, Canongate Books, 2016

O'Sullivan, Suzanne, *It's All in Your Head: Stories from the Frontline of Psychosomatic Illness*, Chatto & Windus, 2015

Articles

Marchant, J, 'You can train your body into thinking it's had medicine', *Mosaic*, 9 February 2016, mosaicscience.com/story/medicine-without-the-medicine-how-to-train-your-immune-system-placebo

Naylor, B, *et al*, 'Reduced glutamate in the medial prefrontal cortex is associated with emotional and cognitive dysregulation in people with chronic pain', *Frontiers in Neurology*, 3 December 2019, doi. org/10.3389/fneur.2019.01110

Norman-Nott, Nell, *et al*, 'Efficacy of the iDBT-Pain skills training intervention to reduce emotional dysregulation and pain intensity in people with chronic pain: protocol for a single-case experimental design with multiple baselines', *BMJ Open*, 14 April 2021, pubmed. ncbi.nlm.nih.gov/33853792

Videos

Harvey, Shannon, *The Connection: Mind Your Body*, www.shannonharvey. com/pages/the-connection-documentary

Websites

The Monash Centre for Consciousness and Contemplative Studies, www. monash.edu/consciousness-contemplative-studies/home

Lynne's **All in the Mind** *episodes*

'The boxer's story', 7 February 2016, www.abc.net.au/radionational/ programs/allinthemind/the-boxers-story/7122138

'Health in body and mind', broadcast 3 March 2019, www.abc.net.
au/radionational/programs/allinthemind/health-in-body-and-
mind/10846326

'Imaginary illness', broadcast 18 October 2015, www.abc.net.au/
radionational/programs/allinthemind/imaginary-illness/6847086

'Linking body and mind', broadcast 19 May 2013, www.abc.net.au/
radionational/programs/allinthemind/new-document/4689134

'Magic happens', broadcast 1 June 2014, www.abc.net.au/radionational/
programs/allinthemind/magic-happens/5478702

'Mind body connection', broadcast 16 November 2014, www.
abc.net.au/radionational/programs/allinthemind/mind-body-
connection/5882006

'Pain on the brain', broadcast 13 March 2016, www.abc.net.au/
radionational/programs/allinthemind/pain-on-the-brain/7232844

'Placebo power', broadcast 8 April 2018, www.abc.net.au/radionational/
programs/allinthemind/placebo-power/9613346

'The science of mind over body', broadcast 20 March 2016, www.abc.net.
au/radionational/programs/allinthemind/the-science-of-mind-over-
body/8139240

'Social lives, genes, and our health', broadcast 18 December 2016, www.
abc.net.au/radionational/programs/allinthemind/social-lives-genes-
and-our-health/8119372

22. Music in Mind

Books

Levitin, Daniel J, *This Is Your Brain on Music: The Science of a Human
Obsession*, Dutton Adult, 2006

Sacks, Oliver, *Musicophilia: Tales of Music and the Brain*, Knopf, 2007

Schulman, Andrew, *Waking the Spirit: A Musician's Journey Healing Body,
Mind, and Soul*, Picador, 2016

Articles

Baker, FA, *et al*, 'A therapeutic songwriting intervention to promote
reconstruction of self-concept and enhance wellbeing following
brain or spinal cord injury: pilot randomised controlled trial',
Clinical Rehabilitation, Vol 33, No 6, doi.org/10.1177/
0269215519831417

Chanda, Mona Lisa, and Levitin, Daniel J, 'The neurochemistry of music:
Evidence for health outcomes', *Trends in Cognitive Sciences*, Vol 17,
No 4 (April 2013), pp 179–193, doi.org/10.1016/j.tics.2013.02.007

Irish, M, *et al*, 'Investigating the enhancing effect of music on autobiographical memory in mild Alzheimer's disease', *Dementia and Geriatric Cognitive Disorders*, Vol 22, No 1, pp 108-120 (2006)

Levitin, Daniel, 'A conversation with Joni Mitchell', 1997, jonimitchell.com/library/originals/jmOriginal_460.pdf

Malcolm, Lynne, 'Tapping into the connections between music and the brain', *ABC Radio National*, 1 April 2016, www.abc.net.au/radionational/programs/allinthemind/tapping-into-the-connections-between-music-and-the-brain/7288328

Tamplin, J, and Baker, FA, (2017), 'Therapeutic singing protocols for addressing acquired and degenerative speech disorders in adults', *Music Therapy Perspectives*, Vol 35, No 2 (2017), pp 113–123, doi.org/10.1093/mtp/mix006

Tamplin, J, *et al*, A theoretical framework and therapeutic songwriting protocol to promote integration of self-concept in people with acquired neurological injuries', *Nordic Journal of Music Therapy*, Vol 25, No 2, (March 2015), pp 111–133, doi.org/10.1080/08098131.2015.1011208

Videos
Alive Inside: A Story of Music and Memory, Music & Memory, 14 April 2012, YouTube, https://bit.ly/3G7Fyca

Websites
Andrew Schulman – Guitarist, Medical Musician, Author, andrewschulmanmusic.com

Music & Memory, musicandmemory.org

Lynne's All in the Mind Episodes
'The medical muso', broadcast 16 April 2017, www.abc.net.au/radionational/programs/allinthemind/the-medical-muso/8430626

'The music in your brain', broadcast 5 January 2014, www.abc.net.au/radionational/programs/allinthemind/the-music-in-your-brain/5132382

'Music of memory', broadcast 31 January 2021, www.abc.net.au/radionational/programs/allinthemind/12943156

'A musical recovery', broadcast 15 June 2014, www.abc.net.au/radionational/programs/allinthemind/musical-recovery/5512228

'Playing for time', broadcast 27 March 2016, www.abc.net.au/radionational/programs/allinthemind/playing-for-time/7271454

'The power of music', broadcast 13 October 2013, www.abc.net.au/radionational/programs/allinthemind/the-power-of-music/4994272

23. The Power of Human Connection

Books

Duerden, Nick, *A Life Less Lonely: What We Can All Do to Lead More Connected, Kinder Lives*, Green Tree, 2018

Ending Loneliness Together, *Ending Loneliness Together in Australia*, White Paper, 2020, www.endingloneliness.com.au/resources/whitepaper/ending-loneliness-together-in-australia

Kelly, Ralph, and Kelly, Kathy, *Too Soon, Too Late: How a Family Turned Its Own Tragedies into a Remarkable Crusade to Keep All Our Children Safe*, Allen & Unwin, 2019

Lim, Michelle, and Australian Psychological Society, *Australian Loneliness Report: A Survey Exploring the Loneliness Levels of Australians and the Impact on Their Health and Wellbeing*, Australian Psychological Society and Swinburne University, 2018, researchbank.swinburne.edu.au/file/c1d9cd16-ddbe-417f-bbc4-3d499e95bdec/1/2018-australian_loneliness_report.pdf

Zaraska, Marta, *Growing Young: How Friendship, Optimism and Kindness Can Help You Live to 100*, Appetite by Random House, 2020

Articles

Bower, Marlee, *et al*, 'In their own words: An Australian community sample's priority concerns regarding mental health in the context of COVID-19', *PLoS ONE*, Vol 17, No 5, e0268824, doi.org/10.1371/journal.pone.0268824

Cole, Steven W, 'The conserved transcriptional response to adversity', *Current Opinions in Behavioural Science*, Vol 28 (August 2019), pp 31–37, pubmed.ncbi.nlm.nih.gov/31592179

Cole, Steven W, *et al*, 'Social regulation of gene expression in human leukocytes', *Genome Biology*, Vol 8, No 9 (2007), R189, pubmed.ncbi.nlm.nih.gov/17854483

Madigan, Sheri, *et al*, 'Association between screen time and children's performance on a developmental screening test', *JAMA Pediatrics*, Vol 173, No 3 (March 2019), pp 244–250, doi.org/10.1001/jamapediatrics.2018.5056

Malcolm, Lynne, 'The brain benefits of looking up from your phone', *ABC News*, 6 July 2019, www.abc.net.au/news/health/2019-07-06/the-brain-benefits-of-looking-up-from-your-phone/11278140

Websites

Happiness & Its Causes, happinessanditscauses.com.au

Imagine Clarity, app.imagineclarity.com

Karuna Shechen: Altruism in Action, karuna-shechen.org/en

Look Up, www.lookup.org.au

OneWave, www.onewaveisallittakes.com

Stay Kind, www.staykind.org

Lynne's All in the Mind *episodes*

'The art of empathy', broadcast 26 August 2018, www.abc.net.au/radionational/programs/allinthemind/the-art-of-empathy/10151614

'Contemplating happiness with Matthieu Ricard', broadcast 2 July 2017, www.abc.net.au/radionational/programs/allinthemind/contemplating-happiness-with-matthieu-ricard/8656456

'End of life care', broadcast 28 May 2017, www.abc.net.au/radionational/programs/allinthemind/end-of-life-care/8547822

'Kindness, and longevity', broadcast 26 July 2020, www.abc.net.au/radionational/programs/allinthemind/kindness-and-longevity/12477978

'Loneliness – A social pain', broadcast 7 April 2019, www.abc.net.au/radionational/programs/allinthemind/loneliness-a-social-pain/10962908

'Look up and connect', broadcast 30 June 2019, www.abc.net.au/radionational/programs/allinthemind/look-up-and-connect/11245692

'A meaningful life', broadcast 26 August 2017, https://www.abc.net.au/radionational/programs/allinthemind/a-meaningful-life/8766686

'Meditation meets neuroscience', broadcast 24 April 2016, www.abc.net.au/radionational/programs/allinthemind/meditation-meets-neuroscience/7339332

'The science of compassion', broadcast 9 December 2012, www.abc.net.au/radionational/programs/allinthemind/the-science-of-compassion/4405514

'Social lives, genes, and our health', broadcast 18 December 2016, www.abc.net.au/radionational/programs/allinthemind/social-lives-genes-and-our-health/8119372

'The story of your brain', broadcast 31 January 2016, www.abc.net.au/radionational/programs/allinthemind/the-story-of-your-brain/7108384

ACKNOWLEDGEMENTS

I wish to thank all those in my life who have supported, encouraged and believed in me, including …

My dear mum, Judith, and dad, Brian, for their love, support and broadmindedness in allowing me to aspire to a lifetime of education and the pursuit of my interests and passions. Though they never articulated it as such, they were determined to give my brother, Peter, and me the educational opportunities they had not had.

My partner, Greg Jewson, for his love and support throughout my career at the ABC, which was often a demanding and all-consuming passion. He believed in me and patiently urged me on during the book-writing process.

My son, Sam Schumacher, and daughter, Eleni Schumacher, who both have shown deep interest in the subject matter and have encouraged me enormously in the writing of this book. I am proud of the wonderful adults they have become, and inspired by their sensitivity, intelligence and creativity.

My brother, Peter; my sister-in-law, Margaret; my niece, Erica; my nephew, Robert; and their partners and families. I'm grateful for their love, support and the good times we share.

My wonderful network of friends, across a range of areas of my life. They are all are so important to me and have enriched me with their support, stimulation and humour.

A very special thanks to Ruth Laxton, my dear friend who's been steadfastly by my side. She enthusiastically offered to read through my draft manuscript. I greatly appreciated her intelligent, sensitive and professional feedback. It was invaluable.

All my friends at the ABC, too many to name. I have grown up with them and I am a better person as a result!

My wonderful work 'family' at ABC RN. They've expanded my mind and helped me to thrive intellectually – and are such a great bunch of creative, talented and stimulating people. With our shared values, it's been an honour to work proudly together in this world-class media outlet.

The ABC Science team, still one of the most highly respected and professional specialist science units in the world. I have learnt an enormous amount from every one of them, both as a science journalist and broadcaster and in my role as an executive producer, which I occupied for a number of years.

While it's not a comprehensive list by any means, I would like to specifically acknowledge and thank the following people who I have worked with in the science unit over the years, for all that they have given and taught me: Linden Woodward, Amanda Armstrong, the late Peter Hunt, the late Alan Saunders, Robyn Williams, Norman Swan, Brigitte Seega, David Fisher, Jonathan Webb, Joel Werner, Natasha Mitchell, Bernie Hobbs, Ann Jones, Olivia Willis, Kylie Andrews, Genelle Weule, Cathy Johnson, Anna Salleh, Tegan Taylor, Nick Kilvert, James Bullen and Carl Smith.

A big thank you, too, to Diane Dean, who was the producer on *All in the Mind* for many years. Together we enjoyed and shared a deep interest in our subject matter and a commitment to make *All in the Mind* the best it could be for our audience. And thanks to the other wonderful producers who worked with me on the program from time to time.

Sana Qadar, now host of *All in the Mind*, who has so ably and beautifully taken the reins and will no doubt carefully guide this important podcast and program into the future.

The talented, dedicated team of ABC audio sound engineers who have worked with me and the production team on creating smooth and top-quality listening experiences for all our programs.

The people at ABC Books and HarperCollins for their experience and wisdom in steering me through the process and for enthusiastically engaging with the content of my book. In particular, thanks to publisher Jude McGee, who encouraged me from the beginning as a new author and kept me on track; copy editor Emma Dowden, who 'got' the content and helped to make it sing; and senior editor at HarperCollins Scott Forbes, who, with his keen eye for detail and balanced sense of judgement, calmly and expertly worked with me and guided me through to the end.

Finally, my deep gratitude goes to the hundreds of academics, scientists, researchers, health professionals and people with lived experience who trusted and shared with me their passion, wisdom and insight into these compelling aspects of human experience. Some of them appear in the book, and many do not, but I know that each of their contributions offers hope and shines light into many people's lives.